Security for Wireless Sensor Networks

Advances in Information Security

Sushil Jajodia
Consulting Editor
Center for Secure Information Systems
George Mason University
Fairfax, VA 22030-4444
email: jajodia@gmu.edu

The goals of the Springer International Series on ADVANCES IN INFORMATION SECURITY are, one, to establish the state of the art of, and set the course for future research in information security and, two, to serve as a central reference source for advanced and timely topics in information security research and development. The scope of this series includes all aspects of computer and network security and related areas such as fault tolerance and software assurance.

ADVANCES IN INFORMATION SECURITY aims to publish thorough and cohesive overviews of specific topics in information security, as well as works that are larger in scope or that contain more detailed background information than can be accommodated in shorter survey articles. The series also serves as a forum for topics that may not have reached a level of maturity to warrant a comprehensive textbook treatment.

Researchers, as well as developers, are encouraged to contact Professor Sushil Jajodia with ideas for books under this series.

Additional titles in the series:

Additional information about this series can be obtained from
http://www.springer.com

Security for Wireless Sensor Networks

by

Donggang Liu
The University of Texas at Arlington, TX
USA

Peng Ning
North Carolina State University, Raleigh, NC
USA

 Springer

Donggang Liu
Dept. Computer Science & Engineering
Univ. of Texas, Arlington
Arlington TX 76019-0015

Peng Ning
North Carolina State University
Dept. of Computer Science
Raleigh NC 27695-8206

Security for Wireless Sensor Networks by Donggang Liu and Peng Ning

e-ISBN-10: 0-387-46781-5
e-ISBN-13: 978-0-387-46781-8

ISBN 978-1-4419-4098-8

Printed on acid-free paper.

9 8 7 6 5 4 3 2 1

springer.com

To my wife Rongfang.
—DL

To my wife Li and my son Daniel.
—PN

Preface

The recent technological advances have made it possible to deploy small, low-power, low-bandwidth, and multi-functional wireless sensor nodes to monitor and report the conditions and events in their local environments. A large collection of these sensor nodes can thus form a wireless sensor network in an ad hoc manner, creating a new type of information systems. Such sensor networks have recently emerged as an important means to study and interact with the physical world and have received a lot of attention due to their wide applications in military and civilian operations such as target tracking and data acquisition. However, in many of these applications, wireless sensor networks could be deployed in hostile environments where there are malicious attacks against the network.

Providing security services in sensor networks, however, turns out to be a very challenging task. First, sensor nodes usually have limited resources such as storage, bandwidth, computation and energy. It is often undesirable to implement expensive algorithms (e.g., frequent public key operations) on sensor nodes. Second, sensor nodes are usually deployed unattended and built without compromise prevention in mind. An attacker can easily capture and compromise a few sensor nodes without being noticed. When sensor nodes are compromised, the attacker can learn all the secrets stored on them and launch a variety of attacks. Thus, any security mechanism for sensor networks has to be resilient to compromised sensor nodes. Third, most sensor applications are based on local computation and communication, while adversaries are usually much more powerful and resourceful than sensor nodes. In many cases, one has to use resource-constrained sensor nodes to deal with very powerful attacks.

This book discusses fundamental security issues in wireless sensor network and presents techniques for the protection of such networks. The purpose of this book is to help both students and professionals to understand fundamental security issues and techniques for wireless sensor network security and prepare them for doing research in this domain. This book could be used as a supplemental textbook covering wireless sensor networks for undergraduate/graduate level security courses with prior knowledge of compute networks, operating systems, probability and statistics, and information security.

This text includes results from recent advances in wireless sensor network security. However, many of these techniques are transitory due to the rapid technological advances in wireless sensor networks. This book is written and organized with special emphasis on basic design principals that may survive current rapid changes in wireless sensor network security.

Donggang Liu
The University of Texas at Arlington
Peng Ning
North Carolina State University

Contents

1

Introduction

This chapter briefly introduces background knowledge about wireless sensor networks as well as the security issues in such network. We will first give a brief overview on wireless sensor networks and the motivations for building trustworthy and resilient sensor networks. We will then present the design challenges for the security mechanisms in sensor networks. Finally, we will introduce the security issues related to wireless sensor networks such as broadcast authentication, key management and secure localization.

1.1 Wireless Sensor Network

Wireless sensor networks have recently emerged as an important means to study and interact with the physical world. A sensor network typically consists of a large number of tiny sensor nodes and possibly a few powerful control nodes (also called *base stations*). Every sensor node has one or a few sensing components to sense conditions (e.g. temperature, humidity, pressure) from its immediate surroundings and a processing and communication component to carry out simple computation on the raw data and communicate with its neighbor nodes. Sensor nodes are usually densely deployed in a large scale and communicate with each other in short distances via wireless links [1]. The control nodes may further process the data collected from the sensor nodes, disseminate control commands to the sensor nodes, and connect the network to a traditional wired network.

Sensor nodes are usually scattered randomly in the field and will form a sensor network after deployment in an ad hoc manner to fulfill certain tasks. There is usually no infrastructure support for sensor networks. As one example, let us look at the battlefield surveillance . In this application, a large number of small sensor nodes are rapidly deployed in a battlefield via airplanes or trucks. After deployment, these sensor nodes are quickly self-organized together to form an ad-hoc network. Each individual sensor node then monitors conditions and activities in its local surroundings and reports its observations

Table 1.1. Characteristic of MICA2 and MICAz motes.

	MICA2	MICAz
Processor	8-bit 7.7MHz ATmega128	8-bit 7.7MHz ATmega128
RAM	4K bytes	4K bytes
ROM	128K bytes	128K bytes
EEPROM	512K bytes	512K bytes
Data Rate	38.4K baud	250k baud
Default Packet Size (under TinyOS[26])	29 bytes	29 bytes
Power Supply	2 AA batteries	2 AA battery

to a central server by communicating with its neighbors. Collecting these observations from sensor nodes allows us to conduct accurate detections on the activities (e.g., possible attacks) of the opposing force and make appropriate decisions and responses in the battlefield.

Obviously, the design of sensor networks requires wireless networking techniques, especially wireless ad hoc networking techniques. However, most traditional wireless networking protocols and algorithms are not suitable for sensor networks. One main challenge of designing a sensor network comes from the resource constraints on sensor nodes. Table 1.1 shows some basic characteristics of typical mote systems such as MICA2 and MICAz [14], which are widely used in current generation of sensor networks.

The wide applications of wireless sensor networks and the challenges in designing such networks have attracted many researchers to develop protocols and algorithms for sensor networks (e.g., [63, 26, 21, 55, 31, 54, 1]). Note that sensor networks may be deployed in hostile environments where enemies may be present. Security becomes a critical issue to make sure the correct operation of sensor networks in many security sensitive scenarios such as military tasks. This book will focus on the security mechanisms of sensor networks in hostile environments, where there are malicious attacks against the network.

1.2 Design Challenges

Security becomes one of the major concerns when there are potential attacks against sensor networks. Many protocols and algorithms (e.g. routing, localization) will not work in hostile environments without security protection. Security services such as authentication and key management are critical to ensure the normal operations of a sensor network in hostile environments. However, some special features of sensor networks make it particularly challenging to provide these security services for sensor networks.

- *Resource constraints* : As shown in Table 1.1, sensor nodes are usually resource constrained, especially energy constrained. Every operation reduces the lifetime of a sensor node. This makes it undesirable to perform expensive operations such as public key cryptography (e.g. RSA [68]) on sensor nodes. For example, a public key is usually 1024 bits (128 bytes) or 2048 bits (256 bytes) long, while a sensor network may only support small size packets as shown in Table 1.1. Though the size of messages can be increased, it is generally not practical to accommodate a long message, since wireless communication is one of the most expensive operations on sensor nodes. In addition, public key operation usually involves many expensive computations (e.g. large integer modular exponentiations).

- *Node compromise*: Different from traditional wireless networks, where each individual node may be physically protected, the large scale of wireless sensor networks makes it impractical to protect or monitor each individual sensor node physically. An attacker may capture or compromise one or a number of sensor nodes without being noticed. If sensor nodes are compromised, the attacker learns all the secrets stored on them and may launch a variety of malicious actions against the network through these compromised nodes. For example, the compromised nodes may discard all important messages in order to hide some critical events from being noticed, or report observations that are significantly different from those observed by non-compromised nodes in order to mislead any decision made based on these data. The result will be even worse if the nodes that provide some critical functions (e.g. data aggregation) are compromised.

  Though using tamper-resistance hardwares may help to protect security sensitive data on sensor nodes, this solution generally increases the cost of an individual sensor node dramatically. An alternative way is to develop security protocols that are *resilient* to node compromise attacks in the sense that even if one or a number of sensor nodes are compromised, the sensor network can still function correctly.

- *Local Computation and Communication versus Global Threats*: The sensor applications in a typical sensor network are usually based on local computation and communication. For example, they may make decisions based on the message exchanged between neighbor nodes. However, adversaries usually are much more powerful and resourceful than sensor nodes, and they usually have a global view of the network (e.g., topology). Thus, we have to use resource-constrained sensor nodes to deal with very powerful attacks.

1.3 Security Issues in Sensor Networks

An important step for protecting sensor networks is the development of fundamental security tools such as broadcast authentication and key management. These fundamental tools provide basic building blocks for us to implement various security mechanisms for sensor networks. On the other hand, sensor network applications are usually supported by many components such as routing and localization. These components clearly have to be protected properly in hostile environments. Thus, in general, the security issues in sensor networks fall into two categories, *fundamental cryptographic tools*, and *security of basic network services*.

This book will discuss techniques for critical security mechanisms in these two categories. In particular, we will focus on broadcast authentication, pairwise key establishment, and secure location discovery.

1.3.1 Broadcast Authentication

Because of the large number of sensor nodes and the broadcast nature of wireless communication, it is usually desirable for base stations to broadcast commands and data to sensor nodes. The authenticity of such commands and data is critical for the normal operation of sensor networks. If convinced to accept forged or modified commands or data, sensor nodes may perform unnecessary or incorrect operations and cannot fulfill the intended purposes of the network. Thus, in hostile environments (e.g., battle field, anti-terrorists operations), it is necessary to enable sensor nodes to authenticate broadcast messages received from base stations.

Providing broadcast authentication in distributed sensor networks is a non-trivial task. On the one hand, public key based digital signatures (e.g., RSA [68]), which are typically used for broadcast authentication in traditional networks, are too expensive to be used in sensor networks, due to the intensive computation involved in signature verification and the resource constraints on sensor nodes. On the other hand, secret key based mechanisms (e.g., HMAC [37]) cannot be directly applied to broadcast authentication, since otherwise a compromised receiver can easily forge any message from the sender.

A protocol named μTESLA [63], which is adapted from a stream authentication protocol called TESLA [61], has been proposed for broadcast authentication in wireless sensor networks. μTESLA employs a chain of authentication keys linked to each other by a pseudo random function [23], which is by definition an one-way function . Each key in the key chain is the image of the next key under the pseudo random function. μTESLA achieves broadcast authentication through delayed disclosure of authentication keys in the key chain. The efficiency of μTESLA is based on the fact that only pseudo random function and secret key based cryptographic operations are needed to authenticate a broadcast message.

The original TESLA uses broadcast to distribute the initial parameters required for broadcast authentication. A digital signature generated by the sender guarantees the authenticity of these parameters. However, due to the low bandwidth of a sensor network and the low computational resources at each sensor node, μTESLA cannot distribute these initial parameters using public key cryptography. Instead, the base station has to unicast the initial parameters to the sensor nodes individually. Such a method certainly cannot scale up to very large sensor networks, which may have thousands of nodes.

Techniques to address this problem are provided in Chapter 2.

1.3.2 Pairwise Key Establishment

Pairwise key establishment is another important fundamental security service. It enables sensor nodes to communicate securely with each other using crypto-graphic techniques. The main problem here is to establish a secure key shared between two communicating sensor nodes. However, due to the resource constraints on sensor nodes, it is not feasible for them to use traditional pairwise key establishment techniques such as public key cryptography and key distribution center (KDC).

Instead of the above two techniques, sensor nodes may establish keys between each other through *key pre-distribution* , where keying materials are pre-distributed to sensor nodes before deployment. As two extreme cases, one may setup a *global* key among the network so that two sensor nodes can establish a key based on this key, or one may assign each sensor node a unique random key with each of the other nodes. However, the former is vulnerable to the compromise of a single node, and the latter introduces huge storage overhead at sensor nodes.

Eschenauer and Gligor proposed a probabilistic key pre-distribution scheme recently for pairwise key establishment [20]. The main idea is to let each sensor node randomly pick a set of keys from a key pool before the deployment so that any two sensor nodes have a certain probability to share at least one common key. Chan et al. further extended this idea and developed two key pre-distribution techniques: a q-composite key pre-distribution scheme and a random pairwise keys scheme [12]. The q-composite key pre-distribution also uses a key pool but requires that two nodes compute a pairwise key from at least q pre-distributed keys that they share. The random pairwise keys scheme randomly picks pairs of sensor nodes and assigns each pair a unique random key. Both schemes improve the security over the basic probabilistic key pre-distribution scheme.

However, the pairwise key establishment problem is still not fully solved. For the basic probabilistic and the q-composite key pre-distribution schemes, as the number of compromised nodes increases, the fraction of affected pairwise keys increases quickly. As a result, a small number of compromised nodes may disclose a large fraction of pairwise keys. Though the random pairwise keys scheme does not suffer from the above security problem, given a memory constraint, the network size is strictly limited by the desired probability that two sensor nodes share a pairwise key, the memory available for keys on sensor nodes, and the number of neighbor nodes that a sensor node can communicate with.

Techniques to address this problem are provided Chapter 3 and Chapter 4.

1.3.3 Security in Localization

Sensors' locations play a critical role in many sensor network applications. Not only do applications such as environment monitoring and target tracking require sensors' location information to fulfill their tasks, but several fundamental techniques developed for wireless sensor networks also require sensor nodes' locations. For example, in geographical routing protocols (e.g., GPSR [35] and GEAR [80]), sensor nodes make routing decisions at least partially based on their own and their neighbors' locations. As another example, in some data-centric storage applications such as GHT [66, 73], storage and retrieval of sensor data highly depend on sensors' locations. Indeed, many sensor network applications will not work without sensors' location information.

A number of location discovery protocols [71, 72, 56, 52, 16, 9, 57, 51, 25] have been proposed for wireless sensor networks in recent years. These protocols share a common feature: they all use some special nodes, called *beacon nodes* , which are assumed to know their own locations (e.g., through GPS receivers or manual configuration). These protocols work in two stages. In the first stage, non-beacon nodes receive radio signals called *beacon signals* from the beacon nodes. The packet carried by a beacon signal, called the *beacon packet* , usually includes the location of the beacon node. The non-beacon nodes then estimate certain measurements (e.g., distance between the beacon and the non-beacon nodes) based on features of the beacon signals (e.g., received signal strength indicator, time of arrival). Such a measurement and the location of the corresponding beacon node collectively is called a *location reference* . In the second stage, a sensor node determines its own location when it has a sufficient number of location references from different beacon nodes. A typical approach is to consider the location references as constraints that a sensor node's location must satisfy, and estimate it by finding a mathematical solution that satisfies these constraints with minimum estimation error.

Despite the recent advances, location discovery for wireless sensor networks in *hostile environments,* where there may be malicious attacks, has been mostly overlooked. The security of location discovery can certainly be enhanced by authentication. Specifically, each beacon packet should be authenticated with a cryptographic key only known to the sender and the intended receivers, and a non-beacon node accepts a beacon signal only when the beacon packet carried by the beacon signal can be authenticated. However, only having authentication does not guarantee the security of location discovery. An attacker may forge beacon packets with keys learned through compromised nodes or replay beacon signals intercepted in different locations. Thus, it is highly desirable to have additional mechanisms to improve the security of location discovery in sensor networks.

Some solutions to address this problem are provided in Chapter 5.

1.4 Orgnization of the Book

The organization of this book is as follows. Chapter 2 presents the multi-level and the tree-based μTESLA broadcast authentication techniques. Chapter 3 discusses the polynomial pool-based key pre-distribution techniques. Chapter 4 presents techniques to improve the performance of pairwise key pre-distribution by using deployment knowledge. Chapter 5 gives details on how to tolerate malicious attacks against location discovery and detect malicious beacon nodes supplying malicious beacon signals. Chapter 6 discusses future research directions on security in wireless sensor networks.

2

Broadcast Authentication

This chapter first introduces the μTESLA broadcast authentication for wireless sensor networks and then presents a *multi-level μTESLA* and a *tree-based μTESLA* to improve the original μTESLA. In multi-level μTESLA protocol, the basic idea is to *predetermine* and *broadcast* the initial parameters required by μTESLA instead of using unicast-based message transmission. In the simplest form, this extension distributes the μTESLA parameters during the initialization of the sensor nodes (e.g., along with the master key shared between each sensor and the base station). To provide more flexibility, especially to prolong the lifetime of μTESLA without requiring a very long key chain, we introduce a multi-level key chain scheme in which the higher-level key chains are used to authenticate the commitments of lower-level ones. To further improve the survivability of the scheme against message loss and Denial of Service (DOS) attacks, we use redundant message transmissions and random selection strategies to deal with the messages that distribute key chain commitments. The resulting scheme removes the requirement of unicast-based initial communication between base station and sensor nodes while keeping the nice properties of μTESLA (e.g., tolerance of message loss, resistance to replay attacks). The experimental results demonstrate that this scheme can tolerate high channel loss rate and is resistant to known DOS attacks to a certain degree.

The tree-based μTESLA can support a large number of senders over a long period of time by using Merkle hash tree [50]. This method has the following advantages over the multi-level μTESLA schemes: (1) It allows broadcast authentication in large sensor networks with a large number of senders, while multi-level μTESLA schemes (as well as the original μTESLA protocol) is not scalable in terms of the number of senders. (2) It is not subject to the DOS attacks against the distribution of μTESLA parameters. In contrast, multi-level μTESLA schemes either consume substantial bandwidth or require significant resources at senders in order to defeat such DOS attacks. To deal with the limited packet payload size in sensor networks, we adopt the idea in [32] and develop a simple method to distribute large messages required for authenticating μTESLA parameters over multiple packets. A nice property of this method is that it allows immediate authentication of the segments of such messages and thus is immune to DOS attacks. We also develop two complementary techniques to revoke broadcast authentication capability from compromised senders: a revocation tree-based scheme and a proactive distribution scheme. The former constructs a Merkle hash tree to revoke compromised senders, while the latter proactively controls the distribution of broadcast authentication capability of each sender to allow the revocation of compromised senders.

2.1 μTESLA Broadcast Authentication

Generally, an asymmetric mechanism, such as public key cryptography, is required to authenticate broadcast messages. Otherwise, a malicious receiver can easily forge any packet from a sender. However, due to the resource constraints at sensor nodes, solutions based on asymmetric cryptography [22, 69, 77] are usually impractical for sensor networks.

One way hash functions have been proposed for authentication in many studies. The use of such functions can be traced back to Lamport [38], which was later implemented as the S/Key one-time password system [24]. Cheung proposed OLSV that uses delayed disclosures of keys by the sender to authenticate link-state routing updates between routers [13]. Anderson et al. used the same technique in their Guy Fawkes protocol to authenticate messages between two parties [2]. Briscoe proposed the FLAMeS protocol [7], and Bergadano et al. presented an authentication protocol for multicast [5]. Both are similar to the OLSV protocol [13]. Canetti et al. proposed to use k different keys to authenticate the multicast messages with k different MAC's for sender authentication [10]; however, this scheme has high communication overhead because of the k MAC's for each message. Perrig introduced a verification efficient signature scheme named BiBa based on one-way hash functions without trapdoors [59]; however, BiBa has high overhead in signature generation and public key distribution.

μTESLA protocol introduces asymmetry by delaying the disclosure of symmetric keys [63]. A sender broadcasts a message with a Message Authentication Code (MAC) generated with a secret key K, which will be disclosed after a certain period of time. When a receiver receives this message, if it can ensure that the packet was sent before the key was disclosed, the receiver can buffer this packet and authenticate it when it receives the corresponding disclosed key. This requires loose time synchronization between the sender and the receivers. To continuously authenticate the broadcast packets, μTESLA divides the time period for broadcasting into multiple time intervals and assigns different keys to different time intervals, as shown in Figure 2.1. All packets broadcasted in a particular time interval are authenticated with the same key assigned to that time interval. For example, packets 1, 2 are authen-

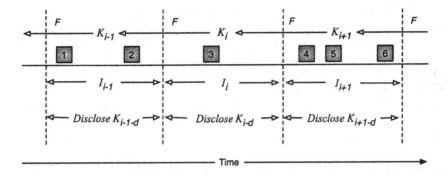

Fig. 2.1. μTESLA protocol.

ticated with K_{i-1}, packet 3 is authenticated with K_i, and packets 4,5,6 are authenticated with K_{i+1}.

To authenticate the broadcast messages, a receiver first authenticates the disclosed keys. μTESLA uses a one-way key chain for this purpose. As shown in Figure 2.1, the sender selects a random value K_n as the last key in the key chain and repeatedly performs a pseudo random function F to compute all the other keys: $K_i = F(K_{i+1}), 0 \leq i \leq n-1$, where the secret key K_i is assigned to the ith time interval. With the pseudo random function F, given K_j in the key chain, anybody can compute all the previous keys $K_i, 0 \leq i \leq j$, but nobody can compute any of the later keys $K_i, j+1 \leq i \leq n$. Thus, with the knowledge of the initial key K_0, which is called the *commitment* of the key chain, the receiver can authenticate any key in the key chain by merely performing pseudo random function operations. When a broadcast message is available in i^{th} time interval, the sender generates MAC for this message with a key derived from K_i and then broadcasts this message along with its MAC and discloses the key K_{i-d} assigned to the time interval I_{i-d}, where d is the disclosure lag of the authentication keys. The sender prefers a long delay in order to make sure that all or most of the receivers can receive its broadcast messages. But, for the receiver, a long delay could result in high storage overhead to buffer the messages.

Each key in the key chain will be disclosed after certain delay. As a result, an attacker can forge a broadcast packet by using the disclosed key. μTESLA uses a security condition to prevent a receiver from accepting any broadcast packet authenticated with a disclosed key. When a receiver receives an incoming broadcast packet at time interval I_i, it checks the security condition $\lfloor (T_c + \Delta - T_0)/T_{int} \rfloor < I_i + d$, where T_c is the local time when the packet is received, T_0 is the start time of the time interval 0, T_{int} is the duration of each time interval, and Δ is the maximum clock difference between the sender and itself. If the security condition is satisfied, i.e., the sender has not disclosed the key K_i yet, the receiver accepts this packet. Otherwise, the receiver simply drops it. When the receiver receives the disclosed key K_i, it can authenticate it with a previously received key K_j by checking whether $K_j = F^{i-j}(K_i)$, and then authenticate the buffered packets that were sent during time interval I_i.

μTESLA is an extension to TESLA [61]. The only difference between TESLA and μTESLA is in their key chain commitment distribution schemes. TESLA uses asymmetric cryptography to bootstrap new receivers, which is impractical for current sensor networks due to its high computation and storage overhead. μTESLA depends on symmetric cryptography with the master key shared between the sender and each receiver to bootstrap the new receivers individually. In this scheme, the receiver first sends a request to the sender, and then the sender replies a packet containing the current time T_c (for time synchronization), a key K_i of the one-way key chain used in a past interval i, the start time T_i of interval i, the duration T_{int} of each time interval and the disclosure lag d.

TESLA was later extended to include an immediate authentication mechanism [62]. The basic idea is to include an image under a pseudo random function of a late message content in an earlier message so that once the earlier message is authenticated, the later message content can be authenticated immediately after it is received. This extension can be applied to μTESLA protocol in the same way.

Perrig et al. proposed to use an earlier key chain to distribute the commitments of the next key chain [60]. Multiple early TESLA packets are used to tolerate packet losses. However, since reliable distribution of later commitment cannot be fully guaranteed, if all the packets used to distribute commitments are lost (e.g., due to temporary network partition), a receiver will not be able to recover the commitment of the later key chain. As a result, the sender and the receivers will have to repeat the costly bootstrap process.

2.2 Multi-Level μTESLA

The major barrier of using μTESLA in large sensor networks lies in its difficulty with distributing the key chain commitments to a large number of sensor nodes. In other words, the method for bootstrapping new receivers in μTESLA does not scale to a large group of new receivers, though it is okay to bootstrap one or a few. The essential reason for this difficulty is the mismatch between the *unicast*-based distribution of key chain commitments and the authentication of *broadcast* messages. That is, the technique is developed for broadcast authentication, but it relies on unicast-based technique to distribute the initial parameters.

In this section, we develop several techniques to extend the capability of μTESLA. The basic idea is to *predetermine* and *broadcast* the key chain commitments instead of unicast-based message transmissions. In the following, we present a series of schemes; each later scheme improves over the previous one by addressing some of its limitations except for Scheme V, which improves over Scheme IV only in special cases where the base station is very resourceful in terms of computational power. The final scheme, a multi-level μTESLA scheme, then has two variations based on schemes IV and V, respectively.

We assume each broadcast message is from the base station to the sensor nodes. Broadcast messages from a sensor node to the sensor network can be handled as suggested in [63]. That is, the sensor node unicasts the message to the base station, which then broadcasts the message to the other sensor nodes. The messages transmitted in a sensor network may reach the destination directly, or may have to be forwarded by some intermediate nodes; however, we do not distinguish between them in our schemes.

For the sake of presentation, we denote the key chain with commitment K_0 as $\langle K_0 \rangle$ throughout this chapter.

2.2.1 Scheme I: Predetermined Key Chain Commitment

A simple solution to bypass the unicast-based distribution of key chain commitments is to predetermine the commitments, the starting times, and other parameters of key chains to the sensor nodes during the initialization of the sensor nodes, possibly along with the master keys shared between the sensor nodes and the base station. (Unlike the master keys, whose confidentiality and integrity are both important, only the integrity of the key chain commitments needs to be ensured.) As a result, all the sensor nodes have the key chain commitments and other necessary parameters once they are initialized, and they are ready to use μTESLA as long as the starting time is passed.

This simple scheme can greatly reduce the overhead involved in distribution of key chain commitments in μTESLA since unicast-based message transmission is no longer required. However, this simple solution also introduces several problems.

First, a key chain in this scheme can only cover a fixed period of time. To cover a long period of time, we need either a long key chain or long time intervals to divide the time period. However, both options may introduce problems. If a long key chain is used, the base station has to allocate a large amount of memory to store the key chain. For example, in our later experiments, the duration of each time interval is 100ms. To cover one day, the base station has to allocate $24 \times 60 \times 60 \times 10 \times 8 = 6,912,000$ bytes memory to store the keys. This may not be desirable in some applications. In addition, the receivers has to perform intensive computation of pseudo random functions if there is a long delay (which covers a large number of time intervals) between broadcast messages in order to authenticate a later disclosed key. Continuing from the previous example, if the time between two consecutive messages received in a sensor is one hour, the sensor has to perform $60 \times 60 \times 10 = 36,000$ pseudo random operations to verify the disclosed key, which may be prohibitive in resource-constrained sensors. If a long interval is used, there will be a long delay before the authentication of a message after it is received, and it requires a larger buffer at each sensor node. Though the extensions to TESLA

[62] can remove the delay in authenticating the data payload and the buffer requirement at the sensor nodes, the messages will have to be buffered longer at the base station.

Second, it is difficult to predict the starting time of a key chain when the sensor nodes are initialized. If the starting time is set too early, the sensor nodes will have to compute a large number of pseudo random functions in order to authenticate the first broadcast message. As shown in the previous example, one hour delay will introduce a huge number of pseudo number operations. In addition, the key chain must be fairly long so that it does not run out before the sensor network's lifetime ends. If the starting time is set too late, messages broadcasted before it cannot be authenticated via μTESLA.

These problems make this simple scheme not a practical one. In the following, we propose several additional techniques so that we not only avoid the problems of unicast-based distribution of key chain commitment but also those of this simple scheme.

2.2.2 Scheme II: Naive Two-Level μTESLA

The essential problem of Scheme I lies in the fact that it is impossible to use both a short key chain and short time intervals to cover a long period of time. This conflict can be mitigated by using multiple levels of key chains. In the following several subsections, we first investigate the special case of two-level key chains to enhance security and robustness, and then extend the results to multi-level key chains in Section 2.2.6.

The two-level key chains consist of a high-level key chain and multiple low-level key chains. The low-level key chains are intended for authenticating broadcast messages while the high-level key chain is used to distribute and authenticate commitments of the low-level key chains. The high-level key chain uses a long enough interval to divide the time line so that it can cover the lifetime of a sensor network without having too many keys. The low-level key chains have short enough intervals so that the delay between the receipt of broadcast messages and the verification of the messages is tolerable.

The lifetime of a sensor network is divided into n_0 (long) intervals of duration Δ_0, denoted as I_1, I_2, ..., and I_{n_0}. The high-level key chain has $n_0 + 1$ elements K_0, K_1, ..., K_{n_0}, which are generated by randomly picking K_{n_0} and computing $K_i = F_0(K_{i+1})$ for $i = 0, 1, ..., n_0 - 1$, where F_0 is a pseudo random function. The key K_i is associated with each time interval I_i. We denote the starting time of I_i as T_i. Thus, the starting time of the high-level key chain is T_1.

Since the duration of the high-level time intervals is usually very long compared with the network delay and clock discrepancies, we choose to disclose a high-level key K_i used for I_i in the following time interval I_{i+1}. Thus, we use the following security condition to check whether the base station has disclosed the key K_i when a sensor node receives a message authenticated with

K_i at time t: $t + \delta_{Max} < T_{i+1}$, where δ_{Max} is the maximum clock discrepancy between the base station and the sensor node.

Each time interval I_i is further divided into n_1 (short) intervals of duration Δ_1, denoted as $I_{i,1}$, $I_{i,2}$, ..., I_{i,n_1}. If needed, the base station generates a low-level key chain for each time interval I_i by randomly picking K_{i,n_1} and computing $K_{i,j} = F_1(K_{i,j+1})$ for $j = 0, 1, ..., n_1 - 1$, where F_1 is a pseudo random function. The key $K_{i,j}$ is intended for authenticating messages broadcasted during the time interval $I_{i,j}$. The starting time of the key chain $\langle K_{i,0} \rangle$ is predetermined as T_i. The disclosure lag for the low-level key chains can be determined in the same way as μTESLA and TESLA [61, 63]. For simplicity, we assume all the low-level key chains use the same disclosure lag d. Further assume that messages broadcasted during $I_{i,j}$ are indexed as (i, j). Thus, the security condition for a message authenticated with $K_{i,j}$ and received at time t is $i' < (i - 1) * n_1 + j + d$, where $i' = \lfloor \frac{t - T_1 + \delta_{Max}}{\Delta_1} \rfloor + 1$, and δ_{Max} is the maximum clock discrepancy between the base station and the sensor node.

When sensor nodes are initialized, their clocks are synchronized with the base station. In addition, the starting time T_1, the commitment K_0 of the high-level key chain, the duration Δ_0 of each high-level time interval, the duration Δ_1 of each low-level time interval, the disclosure lag d for the low-level key chains, and the maximum clock discrepancy δ_{Max} between the base station and the sensor nodes throughout the lifetime of the sensor network are distributed to the sensors.

In order for the sensors to use a low-level key chain $\langle K_{i,0} \rangle$ during the time interval I_i, they must authenticate the commitment $K_{i,0}$ before T_i. To achieve this goal, the base station broadcasts a *commitment distribution message* , denoted as CDM_i, during each time interval I_i. (In the rest of this chapter, we use commitment distribution message and its abbreviation CDM interchangeably.) This message consists of the commitment $K_{i+2,0}$ of the low-level key chain $\langle K_{i+2,0} \rangle$ and the key K_{i-1} in the high-level key chain. Specifically, the base station constructs the CDM_i message as follows:

$CDM_i = i|K_{i+2,0}|MAC_{K_i'}(i|K_{i+2,0})|K_{i-1}$, where "$|$" denotes message concatenation, and K_i' is derived from K_i with a pseudo random function other than F_0 and F_1.

Thus, to use a low-level key chain $\langle K_{i,0} \rangle$ during I_i, the base station needs to generate the key chain during I_{i-2} and distribute $K_{i,0}$ in CDM_{i-2}.

Since the high-level authentication key K_i is disclosed in CDM_{i+1} during the time interval I_{i+1}, each sensor needs to store CDM_i until it receives CDM_{i+1}. Each sensor also stores a key K_j, which is initially K_0. After receiving K_{i-1} in CDM_i, the sensor authenticates it by verifying that $F_1^{i-1-j}(K_{i-1}) = K_j$. Then the sensor replaces the current K_j with K_{i-1}.

Suppose a sensor has received CDM_{i-2}. Upon receiving CDM_{i-1} during I_{i-1}, the sensor can authenticate CDM_{i-2} with K_{i-2} disclosed in CDM_{i-1}, and thus verify $K_{i,0}$. As a result, the sensor can authenticate broadcast mes-

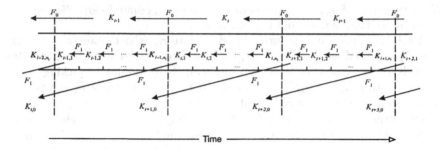

Fig. 2.2. The two levels of key chains in Scheme II. Each key K_i is used for the high-level time interval I_i, and each key $K_{i,j}$ is used for the low-level time interval $I_{i,j}$. F_0 and F_1 are different pseudo random functions. Each commitment $K_{i,0}$ is distributed during the time interval I_{i-2}.

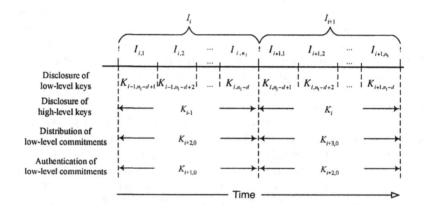

Fig. 2.3. Key disclosure schedule in Scheme II

sages sent by the base station using the μTESLA key chain $\langle K_{i,0} \rangle$ during the high-level time interval I_i.

This scheme uses μTESLA in two different levels. The high-level key chain relies on the initialization phase of the sensor nodes to distribute the key chain commitment, and it only has a single key chain throughout the lifetime of the sensor network. The low-level key chains depend on the high-level key chain to distribute and authenticate the commitments. Figure 2.2 illustrates the two-level key chains, and Figure 2.3 displays the key disclosure schedule for the keys in these key chains.

The two-level key chains scheme mitigates the problem encountered in Scheme I. On the one hand, by having long time intervals, the high-level key chain can cover a long period of time without having a very long key chain.

On the other hand, the low-level key chain has short time intervals so that authentication of broadcast messages does not have to be delayed too much.

The security of this scheme follows directly from the security of μTESLA. Note that the high-level key chain is only used to authenticate the commitment of each low-level key chain. As long as the security condition of each μTESLA key chain is satisfied, the two-level μTESLA has the same degree of security as all the μTESLA instances involved in this scheme. Thus, similar to μTESLA and TESLA, a sensor can detect forged messages by verifying the MAC with the corresponding authentication key once the sensor receives it. In addition, replay attacks can be easily defeated if a sequence number is included in each message.

2.2.3 Scheme III: Fault Tolerant Two-Level μTESLA

Scheme II does not tolerate message losses as well as μTESLA and TESLA. There are two types of message losses: the losses of normal messages, and the losses of CDM messages. Both may cause problems for Scheme II. First, the low-level keys are not entirely chained together. Thus, losses of a key disclosure messages for later keys in a low-level key chain cannot be recovered even if the sensor can receive keys in some later low-level key chains. For example, consider the last key K_{i,n_1} that is used to authenticate the packet in the key chain of time interval I_i. If the CDM message that carries the disclosure of K_{i,n_1} is lost, the sensor then has no way to authenticate this packet. As a result, a sensor may not be able to authenticate a stored message even if it receives some key disclosure messages later. In contrast, with μTESLA a receiver can authenticate a stored message as long as it receives a later key. Second, if CDM_{i-2} does not reach a sensor, the sensor will not be able to use the key chain $\langle K_{i,0} \rangle$ for authentication during the entire time interval I_i, which is usually pretty long (to make the high-level key chain short).

To address the first problem, we propose to further connect the low-level key chains to the high-level one. Specifically, instead of choosing each K_{i,n_1} randomly, we derive each K_{i,n_1} from a high-level key K_{i+1} (which is to be used in the next high-level time interval) through another pseudo random function F_{01}. That is, $K_{i,n_1} = F_{01}(K_{i+1})$. As a result, a sensor can recover any authentication key $K_{i,j}$ as long as it receives a CDM message that discloses $K_{i'}$ with $i' >= i + 1$, even if it does not receive any later low-level key $K_{i,j'}$ with $j' >= j$. Thus, the first problem can be resolved. Figure 2.4 illustrates this idea.

The second problem does not have an ultimate solution; if the base station cannot reach a sensor at all during a time interval I_i, CDM_i will not be delivered to the sensor. However, the impact of temporary communication failures can be reduced by standard fault tolerant approaches.

One possible solution to mitigate the second problem is to include each key chain commitment in multiple CDM messages. For example, we may include each key chain commitment $K_{i,0}$ in l consecutive CDM messages,

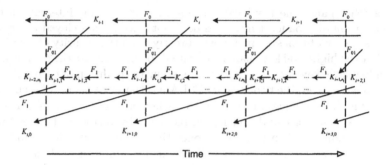

Fig. 2.4. The two levels of key chains in Scheme III. It differs from Figure 2.2 in that each K_{i,n_1} is derived from K_{i+1} using an additional pseudo random function F_{01}.

$CDM_{i-2}, \ldots, CDM_{i-(l+1)}$. As a result, CDM_i includes the key chain commitments $K_{i+2,0}, \ldots, K_{i+1+l,0}$. A sensor can recover and authenticate $K_{i,0}$ if it receives either any two of the above l CDM messages, or one of the l CDM messages and CDM_{i-1}. However, this also increases the size of CDM messages as well as the CDM buffer on sensor nodes. Moreover, the larger a packet is, the more likely it will be lost in wireless communication. Considering the fact that packets in distributed sensor networks usually have limited size (e.g., the payload of each packet in TinyOS [26] is at most 29 bytes), we decide not to go with this solution.

Instead, we propose to have the base station periodically broadcast the CDM message during each time interval. Assuming that the frequency of this broadcast is F, each CDM message is therefore broadcasted $F \times \Delta_0$ times. To simplify the analysis, we assume the probability that a sensor cannot receive a broadcast of a CDM message is p_f. Thus, the probability that a sensor cannot receive any copy of the CDM message is reduced to $p_f^{F \times \Delta_0}$.

Note that even if a sensor cannot receive any CDM message during a time interval I_i, it still has the opportunity to authenticate broadcast messages in time intervals later than I_{i+1}. Not having the CDM message in time interval I_i only prevents a sensor from authenticating broadcast messages during I_{i+1}. As long as the sensor gets a CDM message, it can derive all the low-level keys in the previous time intervals.

By periodically broadcasting CDM messages, Scheme III introduces more overhead than Scheme II. Consider the overhead on the base station, the sensors, and the communication channel respectively. Compared with Scheme II, this scheme does not change the computation of CDM messages in the base station but increases the overhead to transmit CDM messages by $F \times \Delta_0$ times. Base stations in a sensor network are usually much more powerful than the sensor nodes. Thus, the increased overhead on base stations may not be a big problem as long as $F \times \Delta_0$ is reasonable.

The sensors are affected much less than the base station in a benign environment since each sensor only needs to process one CDM message for each time interval. Thus, the sensors have roughly the same overhead as in Scheme II. However, we will show that a sensor has to take a different strategy in a hostile environment in which there are DOS attacks. We will delay the discussion of sensors' overhead until we introduce our counter measures.

This approach increases the overhead in the communication channel by $F \times \Delta_0$ times since the CDM message for each time interval is repeated $F \times \Delta_0$ times. Assume the probability that a sensor cannot receive a CDM message is $p_f = 1/2$ and $F \times \Delta_0 = 10$. Under our simplified assumption, the probability that the sensor cannot receive any of the 10 CDM messages is $p_f^{F \times \Delta_0} < 0.1\%$. Further assume that Δ_0 is 1 minutes, which is quite short as the interval length for the high-level key chain. Thus, there is one CDM message per 6 seconds. Assume the bandwidth is 10 kbps and each CDM packet is 36 bytes = 288 bits, which includes the 29 byte CDM message and the 7 byte packet header as in our experiments (Section 2.2.7). Then the relative communication overhead is $\frac{288}{10240 \times 6} = 0.47\%$. This is certainly optimistic since we assume perfect channel utilization. However, it still shows that Scheme III introduces very reasonable communication overhead in typical sensor networks.

The security of Scheme III is similar to that of Scheme II. The only difference is that each low-level key chain in Scheme III is derived from a high-level key with a pseudo random function F_{01}. Each high-level key is disclosed at least one high-level time interval after the corresponding low-level key chain is used. Thus, as long as the pseudo random function is secure (i.e., it is computationally infeasible to distinguish the output of the pseudo random function from a true random number), Scheme III is equivalent to Scheme II, which does not have F_{01} connecting the two levels of key chains.

One limitation of Scheme III is that if a sensor misses all copies of CDM_i during the time interval I_i, it cannot authenticate any data packets received during I_{i+2} before it receives an authentic K_j, $j > i+2$. (Note that the sensor does not have to receive an authentic CDM message. As long as the sensor can authenticate a high-level key K_j with $j > i+2$, it can derive the low-level keys through the pseudo random functions F_0, F_{01}, and F_1.) Since the earliest high-level key K_j that satisfies $j > i+2$ is K_{i+3}, and K_{i+3} is disclosed during I_{i+4}, the sensor has to buffer the data packets received during I_{i+2} for at least the duration of one high-level time interval.

2.2.4 Scheme IV: DOS-Tolerant Two-Level μTESLA

In Scheme III, the usability of a low-level key chain depends on the authentication of the key chain commitment contained in the corresponding CDM message. A sensor cannot use the low-level key chain $\langle K_{i,0} \rangle$ for authentication before it can authenticate $K_{i,0}$ distributed in CDM_{i-2}. This makes the CDM

messages attractive targets for attackers. An attacker may disrupt the distribution of CDM messages and thus prevent the sensors from authenticating
broadcast messages during the corresponding high-level time intervals. Although the high-level key chain and the low-level ones are chained together,
and such sensors may store the broadcast messages and authenticate them
once they receive a later commitment distribution message, the delay between
the receipt and the authentication of the messages may introduce a problem.
Indeed, an attacker may send a large number of forged messages to exhaust
the sensors' buffer before they can authenticate the buffered messages, and
force them to drop some authentic messages.

The simplest way for an attacker to disrupt the CDM messages is to jam
the communication channel. We may have to resort to techniques such as
frequency hopping if the attacker completely jams the communication channel. This is out of the scope of this chapter. The attacker may also jam the
communication channel only when the CDM messages are being transmitted.
If the attacker can predict the schedule of such messages, it would be much
easier for the attacker to disrupt such message transmissions. Thus, the base
station needs to send the CDM messages randomly or in a pseudo random
manner that cannot be predicted by an attacker who is unaware of the random seed. For simplicity, we assume that the base station sends the CDM
messages randomly.

An attacker may forge commitment distribution messages to confuse the
sensors. If a sensor does not have a copy of the actual CDM_i, it will not be
able to get the correct $K_{i+2,0}$, and cannot use the low-level key chain $\langle K_{i+2,0} \rangle$
during the time interval I_{i+2}.

Consider a CDM message: $CDM_i = i|K_{i+2,0}|MAC_{K_i'}(i|K_{i+2,0})|K_{i-1}$.
Once seeing such a message, the attacker learns i and K_{i-1}. Then the at-
tacker can replace the actual $K_{i+2,0}$ or $MAC_{K_i'}(i|K_{i+2,0})$ with arbitrary values
$K_{i+2,0}'$ or MAC' and forge another message: $CDM_i' = i|K_{i+2,0}'|MAC'|K_{i-1}$.
Assume a sensor has an authentic copy of CDM_{i-1}. The sensor can verify K_{i-1} with K_{i-2} since K_{i-2} is included in CDM_{i-1}. However, the sensor
has no way to verify the authenticity of $K_{i+2,0}'$ or MAC' without the corresponding key, which will be disclosed later. In other words, the sensor cannot
distinguish between the authentic CDM_i messages and those forged by the
attacker. If the sensor does not save an authentic copy of CDM_i during I_i, it
will not be able to get an authenticated $K_{i+2,0}$ even if it receives the authentication key K_i in CDM_{i+1} during I_{i+1}. As a result, the sensor cannot use
the key chain $\langle K_{i+2,0} \rangle$ during I_{i+2}.

One may suggest distributing each $K_{i,0}$ in some earlier time intervals than
I_{i-2}. However, this does not solve the problem. If a sensor does not have an
authentic copy of the CDM message, it can never get the correct $K_{i,0}$. To take
advantage of this, an attacker can simply forge CDM messages as discussed
earlier.

We propose a random selection method to improve the reliable broadcast of
commitment distribution messages. For the CDM_i messages received during

each time interval I_i, each sensor first tries to discard as many forged messages as possible. There is a simple test for a sensor to identify some forged CDM_i messages during I_i. The sensor can verify if $F_0^{i-1-j}(K_{i-1}) = K_j$, where K_{i-1} is the high-level key disclosed in CDM_i and K_j is a previously disclosed high-level key. (Note that such a K_j always exists since the commitment K_0 of the high-level key chain is distributed during the initialization of the sensor nodes.) Messages that fail this test are certainly forged and should be discarded.

The simple test can filter out some forged messages; however, they do not rule out the forged messages discussed earlier. To further improve the possibility that the sensor has an authentic CDM_i message, the base station uses a random selection method to store the CDM_i messages that pass the above test. Our goal is to make the DOS attacks so difficult that the attacker would rather use constant signal jamming instead to attack the sensor network. In other words, we want to prevent the DOS attacks that can be achieved by sending a few packets. Some of the strategies are also applicable to the low-level key chains as well as the (extended) TESLA and μTESLA protocols.

Without loss of generality, we assume that each copy of CDM_i has been weakly authenticated in the time interval I_i by using the aforementioned test.

Single Buffer Random Selection

Let us first look at a simple strategy: *single buffer random selection*. Assume that each sensor node only has one buffer for the CDM message broadcasted in each time interval. In a time interval I_i, each sensor node randomly selects one message from all copies of CDM_i it receives. The key issue here is to make sure all copies of CDM_i have equal probability to be selected. Otherwise, an attacker who knows the protocol may take advantage of the unequal probabilities and make a forged CDM message be selected.

To achieve this goal, for the kth copy of CDM_i a sensor node receives during the time interval I_i, the sensor node saves it in the buffer with probability $1/k$. Thus, a sensor node will save the first copy of CDM_i in the buffer, substitute the second copy for the buffer with probability $1/2$, substitute the third copy for the buffer with probability $1/3$, and so on. It is easy to verify that if a sensor node receives n copies of CDM_i, all copies have the same probability $1/n$ to be kept in the buffer.

The probability that a sensor node has an authentic copy of CDM_i can be estimated as $P(CDM_i) = 1 - p$, where $p = \frac{\#forged\ copies}{\#total\ copies}$. Thus, an attacker has to send as many forged copies as possible to maximize the attack.

Multiple Buffer Random Selection

The single buffer random selection can be easily improved by having additional buffers for the CDM messages. Assume there are m buffers. During each time interval I_i, a sensor node can save the first m copies of CDM_i. For the kth copy with $k > m$, the sensor node keeps it with probability $\frac{m}{k}$. If a copy is to be kept, the sensor node randomly selects one of the m buffers and replaces the corresponding copy. It is easy to verify that if a sensor node receives n copies of CDM_i, all copies have the same probability $\frac{m}{n}$ to be kept in one of the buffers.

During the time interval I_{i+1}, a sensor node can verify if it has an authentic copy of CDM_i once it receives and weakly authenticates a copy of CDM_{i+1}. Specifically, the sensor node uses the key K_i disclosed in CDM_{i+1} to verify the MAC of the buffered copies of CDM_i. Once it authenticates a copy, the sensor node can discard all the other buffered copies.

If a sensor node cannot find an authentic copy of CDM_i after the above verification, it can conclude that all buffered copies of CDM_i are forged and discard all of them. The sensor node then needs to repeat the random selection process for the copies of CDM_{i+1}. Thus, a sensor node needs at most $m + 1$ buffers for CDM messages with this strategy: m buffers for copies of CDM_i, and one buffer for the first weakly authenticated copy of CDM_{i+1}.

It is also easy to see that each sensor node needs to verify the MACs for at most m times. The number of pseudo random function operations required to weakly authenticate the CDM messages depends on the total number of (true and forged) CDM messages a sensor node receives. With m buffer random selection strategy, the probability that a sensor node has an authentic copy of CDM_i can be estimated as $P(CDM_i) = 1 - p^m$, where $p = \frac{\#forged\ copies}{\#total\ copies}$.

Effectiveness of Random Selection

In the rest of this subsection, we perform a further analysis using Markov Chain theory to understand the effectiveness of the random selection strategy. Specifically, we would like to compute the probability that a sensor has an authentic low-level key chain commitment before the key chain is used.

We assume that the base station sends out multiple CDM messages in each high-level time interval so that the probability of all these CDM messages being lost due to lossy channel is negligible. Since our concern is about the availability of an authentic commitment for the low-level key chain before it is used, we consider the state of a sensor only at the end of each high-level time interval.

At the end of each high-level time interval, we use Q_1 to represent that a sensor buffers at least one authentic CDM message in the previous high-level time interval, and Q_2 to represent that a sensor buffers at least one authentic CDM message in the current high-level time interval. We use $\neg Q_1$ (or $\neg Q_2$) to represent that Q_1 (or Q_2) is not true. Thus, with Q_1, Q_2, and their negations, we totally have four combinations, each of which makes one possible state of the sensor. Specifically, state 1 represents $Q_1 \wedge Q_2$, which indicates the sensor has an authentic copy of CDM message in both the previous and the current high-level time interval. Similarly, state 2 represents $Q_1 \wedge \neg Q_2$, state 3 represents $\neg Q_1 \wedge \neg Q_2$, and state 4 represents $\neg Q_1 \wedge Q_2$. A sensor may transit

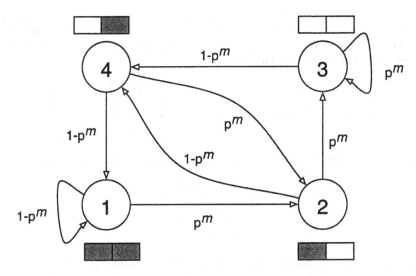

Fig. 2.5. State transition diagram for Scheme IV

from one state to another when the current time moves from the end of one high-level time interval to the end of the next high-level time interval.

Figure 2.5 shows the state transition diagram, which is equivalent to the following transition matrix:

$$
\mathbf{P} = \begin{pmatrix}
1 - p^m & p^m & 0 & 0 \\
0 & 0 & p^m & 1 - p^m \\
0 & 0 & p^m & 1 - p^m \\
1 - p^m & p^m & 0 & 0
\end{pmatrix},
$$

where $p = \frac{\#forged\ copies\ of\ each\ CDM\ message}{\#total\ copies\ of\ each\ CDM\ message}$ and m is the number of buffers for CDM messages in each sensor.

Among the four states, both states 1 and 2 imply that the sensor gets an authentic key chain commitment for the low-level key chain to be used in the next high-level time interval. The reason is as follows: in both states 1 and 2, the sensor already has an authentic CDM message in the previous high-level time interval. Thus, it only needs a disclosed key to authenticate this message. If an attacker wants the DOS attack to be successful, he/she has to ensure the forged CDM messages can be weakly authenticated. As a result, the sensor can obtain a key to authenticate the CDM message distributed in the previous high-level time interval and then obtain an authenticated commitment of the low-level key chain to be used in the next high-level time interval, even if it does not have an authentic copy of the CDM message. Therefore, the overall probability of having an authentic key chain commitment for the next key chain is the sum of the probabilities in state 1 and state 2.

To determine the probability of a sensor being in each state, we need to find the steady state of the above process. Thus, we need to solve the equation $\Pi = \Pi \times \mathbf{P}$, where $\Pi = (\pi_1, \pi_2, \pi_3, \pi_4)$ and π_i represents the probability of the sensor being in state i. That is,

$$(\pi_1, \pi_2, \pi_3, \pi_4) = (\pi_1, \pi_2, \pi_3, \pi_4) \times \begin{pmatrix} 1-p^m & p^m & 0 & 0 \\ 0 & 0 & p^m & 1-p^m \\ 0 & 0 & p^m & 1-p^m \\ 1-p^m & p^m & 0 & 0 \end{pmatrix}.$$

Considering that $\pi_1 + \pi_2 + \pi_3 + \pi_4 = 1$, the solution of the above equation is

$$\begin{cases} \pi_1 = (1-p^m)^2 \\ \pi_2 = p^m(1-p^m) \\ \pi_3 = p^{2m} \\ \pi_4 = p^m(1-p^m). \end{cases}$$

Therefore, the probability that a sensor has an authentic key chain commitment for the next low-level key chain is $P = \pi_1 + \pi_2 = 1 - p^m$. This result shows that the more buffers we have, the more effective this random selection strategy is. Moreover, according to the exponential form of the above formula, having a few more buffers can significantly increase the availability of an authenticated key chain commitment before the key chain is used.

Frequency of CDM Messages

One critical parameter in the proposed technique is the frequency of CDM messages. We describe one way to determine this parameter. Consider a desirable probability P that a sensor has an authenticated copy of a key chain commitment before the key chain is used. Let R_d, R_c and R_a denote the fractions of bandwidth used by data, authentic CDM messages and forged CDM messages, respectively. Assume each message has the same probability p_l of being lost in the communication channel. To simplify the analysis, we assume an attacker uses all available bandwidth to launch a DOS attack. Then we have $R_d + R_c + R_a = 1$. (Note that increasing the transmission of any type of messages will reduce the bandwidth for the other two types of messages. Thus, it is usually difficult in practice to *choose* R_d, R_c, and R_a as desired. Here we consider the relationship among the actual rates as they happen in communication.) To ensure the probability that a sensor has an authentic low-level key chain commitment (before the use of the key chain) is at least P, we have

$$1 - \left(\frac{R_a \times (1 - p_l)}{R_c \times (1 - p_l) + R_a \times (1 - p_l)} \right)^m \geq P.$$

This implies

$$R_a \leq \frac{\sqrt[m]{1 - P}}{1 - \sqrt[m]{1 - P}} \times R_c.$$

Together with $R_d + R_c + R_a = 1$, we have

$$R_c \geq (1 - R_d)(1 - \sqrt[m]{1 - P}). \tag{2.1}$$

Equation 2.1 presents a way to determine the frequency of CDM messages to mitigate severe DOS attacks that use all available bandwidth to prevent the distribution and authentication of low-level key chain commitments. In other words, if we can determine the number m of CDM buffers based on resources on sensors, the fraction R_d of bandwidth for data packets based on the expected application behaviors, the probability P of a sensor authenticating a low-level key chain commitment before the key chain is used based on the expected security performance under severe DOS attacks, we can compute R_c and then determine the frequency of CDM messages. Moreover, we may examine different choices of these parameters and make a trade-off most suitable for the sensor networks.

Figure 2.6 shows the fraction of bandwidth required for CDM messages for different combinations of R_d and m given $P = 0.9$. We can see that the bandwidth required for CDM messages in order to ensure $P = 0.9$ is substantially more than that required to deal with message losses. For example, as shown in Figure 2.6(a), when there are few data packets and each sensor has only 10 buffers for CDM messages, about 20% of the bandwidth must be used for CDM messages in order to ensure 90% authentication rate for low-level key chain comments when there are severe DOS attacks. This is understandable since under such circumstances the sensor network is facing aggressive attackers that try everything possible to disrupt the normal operations of the network.

It is also shown in Figure 2.6(b) that the increase in the number of CDM buffers can significantly reduce the requirement for CDM messages. As shown in Figure 2.6(b), when each sensor has 40 CDM buffers, less than 5% of the bandwidth is required for CDM messages. In addition, the shape of the curves in Figure 2.6(b) also shows that the smaller m is, the more effective an increase in m is.

Figure 2.6(a) further shows that the increase in data rate results in the decrease in the fraction of bandwidth required for CDM messages. This is because when the data consume more bandwidth, there is less bandwidth for the DOS attacks, and in effect the requirement for CDM messages is also reduced.

It is worth noting that the fractions for data and CDM messages are the *actual* fractions of CDM messages that the sensors receive, not the fractions *planned* by the base station. A message scheduled for transmission by the base station is not guaranteed to be transmitted if the DOS attack consumes too much bandwidth. Nevertheless, the above analysis provides a target frequency of CDM messages, and the base station can adaptively change its transmission strategy to meet this target.

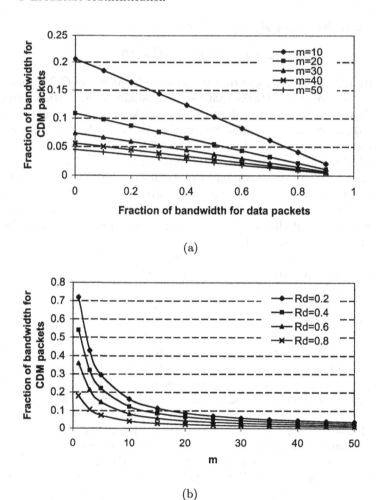

(a)

(b)

Fig. 2.6. Bandwidth required for CDM messages to ensure 90% of low-level key chain commitments are authenticated before the key chains are used.

2.2.5 Scheme V: DOS-Resistant Two-Level μTESLA

Scheme IV can be further improved if the base station has enough computational and storage resources. Indeed, when at least one copy of each CDM message can reach the sensors, we can completely defeat the aforementioned DOS attack without the random selection mechanism.

The solution can be considered a variation of the immediate authentication extension to TESLA [62]. The idea is to include in CDM_i the image $H(CDM_{i+1})$ for each i, where H is a pseudo random function. As a result, if a sensor can authenticate CDM_i, it can get authentic $H(CDM_{i+1})$ and then authenticate CDM_{i+1} when it is received. Specifically, the base station constructs CDM_i for the high-level time interval I_i as follows:

$$CDM_i = i|K_{i+1,0}|H(CDM_{i+1})|MAC_{K_i'}(i|K_{i+1,0}|H(CDM_{i+1}))|K_{i-1},$$

where "|" denotes message concatenation, H is a pseudo random function other than F_0 and F_1, and K_i' is derived from K_i with a pseudo random function other than H, F_0 and F_1.

Suppose a sensor has received CDM_i. Upon receiving CDM_{i+1}, the sensor can authenticate CDM_i with K_i disclosed in CDM_{i+1}. Then the sensor can immediately authenticate CDM_{i+1} by verifying that applying H to CDM_{i+1} results in the same $H(CDM_{i+1})$ included in CDM_i. As a result, the sensor can authenticate a commitment distribution message immediately after receiving it.

Alternatively, if $H(CDM_1)$ is pre-distributed before deployment, the sensor can immediately authenticate CDM_1 when receiving it, and then use $H(CDM_2)$ included in CDM_1 to authenticate CDM_2, and so on. One may observe that in this case, a sensor does not use the disclosed high-level keys in CDM messages directly. However, including such keys in CDM messages is still useful. Indeed, when a sensor fails to receive or keep an authentic CDM message, it can use the random selection mechanism and the approach described in the previous paragraph to recover from the failure.

The cost, however, is that the base station has to compute the CDM messages in the reverse order. That is, in order to include $H(CDM_{i+1})$ in CDM_i, the base station has to have CDM_{i+1}, which implies that it also needs CDM_{i+2}, and so on. Therefore, the base station needs to compute both the high-level and the low-level key chains completely to get the commitments of these key chains and construct all the CDM messages in the reverse order before the distribution of the first one of them. (Note that in Scheme IV, the base station only needs to compute the high-level key chain, not all the low-level ones during initialization. The base station may delay the computation of a low-level key chain until it needs to distribute the commitment of that key chain.)

This imposes additional computation during the initialization phase. Assume that all the key chains have 1,000 keys. The base station needs to perform about 1,001,000 pseudo random function operations to generate all the key chain commitments, and 1,000 pseudo random function operations and 1,000 MAC operations to generate all the CDM messages. Due to the efficiency of pseudo random functions, such computation is still practical if the base station is relatively resourceful. For example, using MD5 as the pseudo random function, a modern PDA can finish the above computation in several seconds. Moreover, the base station does not have to save the low-level key chains. Indeed, to reduce the storage overhead, the base station may compute a low-level key chain (again) when the key chain is needed. Thus, the base station only needs to store the high-level key chain and the MACs of all the

CDM messages. Further assume both the authentication key and the image of a pseudo random function are 8 bytes. To continue the earlier example, the base station needs $(8 + 8) \times 1,000 = 16,000$ bytes to store the high-level key chain and the MACs.

The immediate authentication of CDM_i depends on the successful receipt of CDM_{i-1}. However, if a sensor cannot receive an authentic CDM_i due to communication failure or an attacker's active disruption, the sensor has to fall back to the techniques introduced in Scheme IV (i.e., the random selection strategies). This implies that the base station still needs to distribute CDM messages multiple times in a random manner. The combination of these techniques is straightforward; we do not discuss it further in this chapter.

Now let us assess how difficult it is for a sensor to recover if it fails to receive an authentic CDM message. We assume an attacker will launch a DOS attack to deter this recovery. To recover from the failure, the sensor has to buffer an authentic CDM message by the end of a later high-level time interval and then authenticate this message. For example, suppose a sensor buffers an authentic CDM_{i+j}. If it receives a disclosed key in interval I_{i+j+1}, it can authenticate CDM_{i+j} immediately and gets $H(CDM_{i+j+1})$. The sensor then recovers from the failure. Thus, if a sensor fails to receive an authentic CDM_i, the probability that it recovers from this failure within the next l high-level time intervals is $1 - p^{m \times l}$, where $p = \frac{\#forged\ copies\ of\ each\ CDM\ message}{\#total\ copies\ of\ each\ CDM\ message}$ and m is the number of buffers for CDM messages.

It is sensible to dynamically manage CDM buffers in sensors in this scheme. There are three cases: (1) During normal operations, each sensor only needs one buffer to save an authenticated CDM message during each high-level time interval; (2) When a sensor tries to recover from communication failures, it needs a relatively small number of CDM buffers to tolerate communication failures, as discussed in Section 2.2.3; (3) When a sensor tries to recover from a loss of authentic CDM messages under severe DOS attacks, the sensor needs as many buffers as possible to increase its chance of recovery. Once a sensor recovers an authentic CDM message, it can fall back to only one CDM buffer since it can authenticate the next CDM message once the message is received. This requires that each sensor be able to detect the presence of DOS attacks. Fortunately, this can be done easily with high precision. If most buffered CDM messages are forged, there must be a DOS attack.

The base station needs to broadcast each CDM message multiple times to mitigate communication failures and help sensors recover from failures under potential DOS attacks. The frequency of CDM messages required in this scheme can be determined in a similar way to Scheme IV. However, a sensor in this scheme only needs a large number of CDM buffers temporarily during recovery. Moreover, a sensor only needs to recover one authentic CDM message in order to go back to normal operations, and the sensor may recover over several high-level time intervals. Indeed, if we allow a sensor to recover from such a failure over l high-level time intervals, we can get the following equation by using the same process to derive Equation 2.1:

$$R_c \geq (1 - R_d)(1 - \sqrt[m \cdot l]{1 - P}), \qquad (2.2)$$

where R_c is the fraction of bandwidth required for CDM messages, R_d is the fraction of bandwidth used by data packets, m is the number of buffers for CDM messages, and P is the desired probability to recover from the failure over the next l high-level time intervals. It is easy to see that R_c decreases when m and l increase. Thus, the bandwidth required for CDM messages can be much less than in Scheme IV.

Since the probability that a sensor fails to receive an authentic CDM message is unknown, it is not possible to derive the probability that the sensor has an authentic low-level key chain commitment before the key chain is used. Nevertheless, this probability can be easily computed in the same way as in Section 2.2.4 if the aforementioned information is available.

From the above analysis, we can see that this scheme introduces additional computation requirement before deployment, though it can defeat the DOS attacks when at least one copy of each CDM message reaches the sensors. Fortunately, such computation is affordable if the base station is relatively resourceful. It is also possible to perform such computation on powerful machines and then download the result to the base station before deployment. In addition, the communication overhead and the sensor storage overhead in this scheme is potentially much less than that in Scheme IV, as discussed earlier. Thus, when the required computational resources are available (on either the base station or some other machines), Scheme V is more desirable. Otherwise, Scheme IV could be used to mitigate the DOS attacks.

2.2.6 Scheme VI: Multi-Level μTESLA

Both Scheme IV and Scheme V can be extended to M-level key chain schemes. The M-level key chains are arranged from level 0 to level $M-1$ from top down. The keys in the $(M - 1)$-level key chains are used for authenticating data packets. Each higher-level key chain is used to distribute the commitments of the immediately lower-level key chains. Only the last key of the top-level (level 0) key chain needs to be selected randomly; all the other keys in the top-level key chain are generated from this key, and all the key chains in level i, $1 \leq i \leq M - 1$, are generated from the keys in level $i - 1$, in the same way that the low-level key chains are generated from the high-level keys in the two-level key chain schemes. For security concerns, we need a family of pseudo random functions. The pseudo random function for each level and between adjacent levels should be different from each other. Such a family of pseudo random functions has been proposed in [61].

The benefit of having multi-level key chains is that it is more flexible in providing short key chains with short delays in authenticating data packets, compared with the two-level key chain schemes. As a result, a multi-level μTESLA scheme can scale up to cover a long period of time. In practice, a three-level scheme is usually sufficient to cover the lifetime of a sensor network.

For example, if the duration of a lowest-level time interval is 100ms, and each key chain has 1,000 keys, then a three-level scheme can cover a period of 10^8 seconds, which is over three years. In the following, we still present our techniques as generic multi-level key chains schemes for the sake of generality.

In addition to multi-level μTESLA schemes directly extended from schemes IV and V, we can combine them into a hybrid scheme to achieve a trade-off between pre-computation and operational overheads. Thus, we have three variations of multi-level μTESLA schemes. The first variation, which is named *DOS-tolerant multi-level μTESLA*, is extended from Scheme IV and is suitable for sensor networks where the base station is not very resourceful. The second variation, which is named *DOS-resistant multi-level μTESLA*, is extended from Scheme V. This variation is suitable for sensor networks with relatively short lifetime and relatively powerful base stations. The third variation, which is named *hybrid multi-level μTESLA*, is a trade-off between the above two variations. It sacrifices certain immediate authentication capability in exchange for less pre-computation requirement.

In the following, we describe and analyze these variations, respectively.

Variation I: DOS-Tolerant Multi-Level μTESLA

This variation of multi-level μTESLA scheme is a direct extension to Scheme IV. Each CDM message has the same format as in Scheme IV, and each sensor uses the multiple buffer random selection mechanism to save CDM messages. The only difference is that this variation may have more than two key chain levels.

Compared with Scheme IV, this variation is not more vulnerable to DOS attacks. The success of the DOS attacks depends on the percentage of forged CDM messages and the buffer capacity in sensor nodes. As long as the base station maintains a certain authentic CDM message rate, this variation will not have a higher percentage of forged CDM messages than Scheme IV. The base station can further piggy-back the CDM messages for different levels of key chains so as to reduce the communication cost.

Having more levels of key chains does increase the overhead at both the base station and the sensor nodes. This variation requires the base station to maintain one active key chain at each level. Because of the available resource in typical bases stations, this overhead is usually tolerable. Similarly, sensor nodes have to maintain more buffers for the key chain commitments as well as CDM messages in different key chain levels. This is usually not desirable because of the resource constraints in sensors. In addition, the more levels we have, the more bandwidth is required to transmit the CDM messages. Thus, we should use as few levels as possible to cover the lifetime of a sensor network.

Now let us consider how to configure the frequency of CDM messages in DOS-tolerant multi-level μTESLA. To increase the chance of success, the attacker may target a particular key chain level instead of attacking all

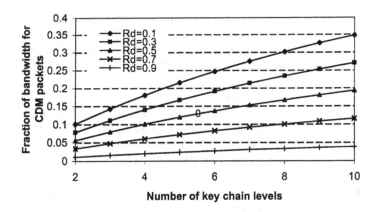

Fig. 2.7. Bandwidth for CDM messages v.s. number of key chain levels. Assume the number of CDM buffers in each key chain level is $m = 40$.

levels simultaneously. Further assume that the base station sends out the CDM messages of each key chain level in the same frequency, and the buffer in each sensor can accommodate m (authentic and/or forged) copies of a CDM message. Thus, for DOS-tolerant M-level μTESLA, Equation 2.1 can be generalized to

$$R_c \geq \frac{(M-1)(1-R_d)(1-\sqrt[m]{1-P})}{(M-1)(1-\sqrt[m]{1-P}) + \sqrt[m]{1-P}}, \tag{2.3}$$

where R_c is the fraction of bandwidth required for CDM messages in all key chain levels, and R_d is the fraction of bandwidth used for data packets, m is the number of CDM buffers in each key chain level, and P is the desired probability that a sensor has an authenticated key chain commitment before the key chain is used.

We may still use the approach in Section 2.2.4 to determine the frequency of CDM messages in order to maintain broadcast authentication service when the network is under severe DOS attacks. Figure 2.7 shows the required fraction of bandwidth for CDM messages to guarantee that each sensor has the probability $P = 0.9$ to have an authenticated low-level key chain commitment before the key chain is used. It is easy to see that the addition of more key chain levels does introduce additional communication overhead. Similar to Figure 2.6, Figure 2.7 shows a smaller fraction of bandwidth required for CDM messages when the data rate is higher. As discussed earlier, the increase in data rate consumes more bandwidth for data and leaves less bandwidth for forged CDM messages. As a result, the requirement for CDM messages is also reduced.

In the following, we give an analysis of the overheads introduced by the DOS-tolerant multi-Level μTESLA scheme. For simplicity, we assume there are totally M levels in our scheme and L keys in each key chain. Thus, if the duration of each lowest-level time interval (level $M - 1$) is Δ, the duration of each level i time interval is $\Delta_i = \Delta \times L^{M-i-1}$, and the maximum lifetime of the scheme is $\Delta \times L^M$.

The storage overhead in sensors is mainly due to the buffer of CDM messages. Each sensor has to buffer weakly authenticated CDM messages for the top $M - 1$ levels. Assuming a sensor uses m CDM buffers, this totally requires about $m \cdot (M - 1)$ buffers. (Note that for each CDM message, only the disclosed key chain commitment and the MAC need to be stored.) In addition, each sensor needs to store 1 most recently authenticated key for level 0 key chain and 3 most recently authenticated keys for each of the other levels (one for the previous key chain because it is possible that the sensor receives a packet which discloses a key in the previous key chain, another for the current key chain, and a third for the next key chain). Thus, each sensor needs to store $3M - 2$ more keys.

A base station only needs to keep the current key chain for each level, which occupies at most $M \times L$ storage space in total. This is because a lower-level key chain can be generated directly from a key in its adjacent upper-level key chain, and the length of key chain in our technique can be short enough to allow computation of a key chain in real time. In contrast, in the original μTESLA scheme [63], the base station has to pre-compute and store L^M keys to cover the same period of time as in our scheme.

Consider the communication overhead due to the CDM messages. In order to mitigate severe DOS attacks, the base station has to use a fair amount of bandwidth to broadcast CDM messages, as indicated by Equation 2.3. For example, Figure 2.7 shows that when the fraction of bandwidth for data packets is 0.1, the number of key chain levels is 3, and each sensor has 40 buffers for each CDM message, the base station needs about 15% of the bandwidth for CDM messages.

The computational overhead in sensors is mainly due to the authentication of disclosed keys and MACs. A sensor's computation for data packets is dependent on the number of data packets the sensor receives. However, a sensor's computation for CDM packets is bounded by the number m of CDM buffers since the sensor has at most m copies of each CDM message, and it can stop once it authenticates a copy.

As discussed earlier, in the original μTESLA protocol, if there is a long delay between the receipts of two data packets, a sensor has to perform a large number of pseudo random functions in order to authenticate the key disclosed in the packet. In the worst case, it has to perform about L^M pseudo random functions if it only receives the first and the last packets. In contrast, with the DOS-tolerant multi-level μTESLA scheme, such a sensor needs to perform at most $M \times L$ pseudo random functions. In general, if a sensor does not receive packets for n_l lowest-level time intervals, the number of pseudo random functions that it needs to perform in order to authenticate a key received later never exceeds $L \times \log_L(n_l)$.

It appears that the overheads in this scheme, especially the communication overhead and the storage overhead in sensors, are not negligible. In the following, we introduce the second variation of multi-level μTESLA scheme that is more efficient in terms of communication overhead and storage overhead in sensors.

Variation II: DOS-Resistant Multi-Level μTESLA

The DOS-resistant multi-level μTESLA scheme is extended directly from scheme V. There are multiple key chain levels, with lower-level key chains generated from keys in the immediately higher-level key chains. There are multiple key chains in all levels except for level 0. Among these levels, only level $M - 1$ is used to authenticate data packets; all the other levels are used to distribute the key chain commitments in the immediately lower-level. Each CDM message consists of the image of the next CDM message under a pseudo random function. In level i, $0 < i < M - 1$, the last CDM message in an earlier key chain contains the image of the first CDM message in the immediately next key chain. As a result, the end of a key chain does not interrupt the immediate authentication of later CDM messages in the same level.

Similar to its two-level counter part, this scheme requires pre-computation to generate all the key chains in each level and all the CDM messages. This computation cost could be prohibitive if the lifetime of a sensor network is very long. However, it may be tolerable for relatively short-lived sensor networks. For example, consider a three-level scheme with 100 keys in each key chain and 100ms lowest-level time intervals. Such a scheme can cover 10^5 seconds, which is about 27 hours. The pre-computation required to initialize the scheme consists of 1,010,100 pseudo random functions to generate all the key chains, and 10,100 pseudo random functions to generate all the CDM messages. Such computation can be finished in several seconds on a modern PC or PDA. Thus, the pre-computation can be either performed on base stations directly, or performed on a regular PC and then downloaded to the base station.

The base station does not have to store all these values due to the low cost involved in computing pseudo random functions. To continue the above example, the base station may simply store the keys for the active key chain of each level and the images of CDM messages under pseudo random functions. Assume that both a key and an image of a pseudo random function takes 8 bytes. Then the base station only needs to save about $8 \times 300 + 8 \times 10,100 \approx$ 82 KBytes.

In general, for a DOS-resistant M-level μTESLA scheme, where each key chain consists of L keys, a base station needs to pre-compute $L+L^2+...+L^M = \frac{L^{M+1}-L}{L-1}$ keys and $L+L^2+...+L^{M-1} = \frac{L^M-L}{L-1}$ CDM messages, respectively. In addition, the base station needs to store $M \times L$ keys and $\frac{L^M - 1}{L - 1}$ CDM images, respectively. Additional trade-off is possible to reduce the storage requirement (by not saving but computing some CDM images when they are needed) if the base station does not have space for all these keys and CDM images.

This scheme inherits the advantage of its two-level counter part. That is, a sensor can get an authenticated key chain commitment as long as it receives one copy of the corresponding CDM message. As we discussed in Section 2.2.5, this property substantially reduces the communication overhead introduced by CDM messages since the base station only needs to send enough copies of a CDM message to make sure the sensors have a high probability to receive CDM messages during normal operations, and have a high probability to recover from failures over a period of time when the sensors are under DOS attacks. Specifically, if we would like a sensor to recover from a failure of receiving a CDM message within l time intervals (in the same level), by using the same process to get Equation 2.3, we have the following equation:

$$R_c \geq \frac{(M-1)(1-R_d)(1-\sqrt[m \cdot l]{1-P})}{(M-1)(1-\sqrt[m \cdot l]{1-P})+\sqrt[m \cdot l]{1-P}}, \tag{2.4}$$

where R_c is the fraction of bandwidth required for CDM messages in all key chain levels, and R_d is the fraction of bandwidth used for data packets, m is the number of CDM buffers in each key chain level, and P is the desired probability that a sensor recovers from the failure over the next l time intervals. It is easy to verify that when m and l increase, the right hand side of Equation 2.4 decreases, and so does the requirement for R_c. Moreover, a sensor may use dynamic buffer management as discussed in

Section 2.2.5 to arrange buffers for CDM messages. Though a CDM message in this scheme is slightly larger than that in variation I (by one pseudo random function image per CDM message), the frequency of CDM messages can be reduced substantially. Thus, the overall storage requirement in sensors can be much less than that in Variation I.

The computational overhead in a sensor is not as clear as in Variation I. In Variation I, the number of authentication operations a sensor needs to perform is bounded by the number of CDM buffers. In contrast, in this scheme, a sensor may only need to authenticate one copy of CDM message if the first received message is authentic, but it may also have to authenticate every received copy of a CDM message if no copy is authentic in the worst case.

The limitation of this variation is its scalability. It is easy to see that the pre-computation cost is linear to the number of lowest-level time intervals. Consider a long-lived sensor network that requires a 3-level key chains scheme, where each key chain consists of 1,000 keys and the duration of each lowest-level time interval is 10ms. The lifetime of this scheme is 10^7 seconds, which is about 116 days. Using 3-level key chains implies that the base station needs to pre-compute about 1,001,001,000 pseudo random functions to compute the key chains and another 1,001,000 pseudo random functions to compute the images of CDM messages. In addition, the base station needs to store about 3000 keys and 1,001,000 images of pseudo random functions, which take about 8 MBytes memory. Though this is still feasible for typical PCs and workstations, it may be too expensive for base stations that are not very resourceful.

Variation III: Hybrid Multi-Level μTESLA

Variation III is essentially a trade-off between the first two variations. To make the techniques in Variation II practical for low-end base stations, we reduce the pre-computation and storage overheads by sacrificing certain immediate authentication capability. Specifically, we limit the pre-computed CDM messages to the active key chain being used in each level. For a given key chain in a particular level, the base station computes the images of the CDM messages (under the pseudo random function H) only when the first key is needed for authentication, and this computation does not go beyond this key chain in this level. As a result, the CDM message authenticated with the last key in a key chain will not include the image of the next CDM message in the same level, because this information is not available yet. The base station may simply set this field as NULL. For the first key chain in each level i, where $0 \leq i \leq M-1$, the image of the first CDM message can be distributed during the initialization phase.

The behavior of a sensor is still very simple. If the sensor has an authentic image of the next CDM message in a certain level, it can authenticate the next CDM message immediately after receiving it. Otherwise, the sensor simply uses the random selection strategy to buffer the weakly authenticated copies. To increase the chance that the sensors receive an authentic image of the first CDM message for a key chain, the base station may also broadcast it in data packets.

Such a method reduces the computation and storage requirements significantly compared with Variation II. For an M-level μTESLA with L keys in each key chain, the base station only needs to pre-compute around $M \cdot L$ pseudo random functions and store $(M-1) \cdot L$ images of CDM messages. In the earlier example with 3-level key chains and 1,000 keys per key chain, the base station only needs to compute about 3,000 (instead of 1,001,001,000 in Variation II) pseudo random operations during initialization and store 2,000 (instead of 1,001,000 in Variation II) CDM images.

An obvious weak point of this multi-level μTESLA scheme is the handover of two consecutive key chains in the same level. Consider two consecutive key chains in level i, where $i < M-1$. These key chains are used to distribute CDM messages for the immediately lower-level key chains. For all the keys except for the last one in each key chain, the corresponding CDM messages include an image of the next CDM message, which enables a sensor to authenticate the next CDM message immediately after receiving it. However, the last CDM message corresponding to the earlier key chain does not have an image of the first CDM message corresponding to the later key chain, as discussed earlier. Thus, the first CDM message of the later key chain cannot be authenticated immediately after it is received, though the commitment of this key chain can be authenticated with the immediately upper-level CDM message. As a result, a sensor has to wait for the next CDM message to disclose the corresponding μTESLA key in order to authenticate the first CDM message.

An attacker may take advantage of this opportunity to launch DOS attacks. However, this scheme will not perform worse than Variation I since each sensor can always fall back to the random selection mechanism to mitigate the impact of such an attack. In addition to the dynamic buffer management discussed in Section 2.2.5, the base station can also use an adaptive method to determine the frequency of CDM messages to improve the resistance against DOS attacks without substantially increasing the communication overhead. That is, the base station may use a low frequency to send out CDM messages corresponding to later intervals in a key chain and use a high frequency for the early ones. The analysis performed for Variation I to decide the desirable frequency of CDM messages is also applicable to Variation III.

Although having less overhead than Variation II, Variation III introduces more overheads into base stations than Variation I. In addition to computing a key chain before using it, a base station using this variation has to compute all the corresponding CDM messages since each earlier CDM message includes the image of the immediately following CDM message. The storage overhead in the base station in this scheme is also higher than that in Variation I, due to the storage of these CDM messages.

Variation III introduces lower overheads in sensors than Variation I, but has higher overheads than Variation II. In normal situations when a sensor has an authenticated image of the following CDM message, it only needs to save one copy of that CDM message. A sensor's computation and storage overheads are the same as in Variation II. During the handover of two key chains (in the same level), a sensor needs to increase the number of CDM buffers to mitigate potential DOS attacks. This is similar to Variation I. However, unlike in Variation I, a sensor using Variation III can recover to the above normal situation once it authenticates one CDM message. This is essentially the same as recovering from failures (to receive an authentic CDM message) in Variation II. As discussed earlier, the storage overhead in sensors is much smaller than that in Variation I when the sensors are allowed to recover over several time intervals. But such overheads in a sensor using Variation III are higher than in Variation II since such recovery processes are "scheduled" in addition to those due to failures.

A sensor using Variation III may use an adaptive approach to save CDM messages during handover of key chains. Specifically, a sensor may just save a few (or even a single copy) of the first CDM message corresponding to a new key chain. When the next CDM message arrives, the sensor can then decide whether there is an on-going DOS attack by attempting to authenticate the earlier CDM message. If the earlier CDM message is authenticated, the sensor can continue to authenticate later CDM messages with the corresponding image; otherwise, the sensor can determine that there is a DOS attack and adaptively increase the number of CDM buffers.

Consider the communication overhead in Variation III introduced by CDM messages. We can use Equation 2.4 to determine the frequency of CDM messages given the fraction of bandwidth used by data packets, the number M of key chain levels, the number m of CDM buffers in each sensor, and the probability P that a sensor recovers from a failure (or gets the first authenticated CDM message for a key chain) over l time intervals. The base station may increase the frequency of the first several CDM messages in a key chain based on Equation 2.3 to increase their probability of being authenticated by sensors. Thus, the communication overhead in Variation III is between those of Variation I and Variation II.

Among these variations, Variation II has a distinctive advantage over the other two variations. Indeed, Variation II can substantially reduce the impact of DOS attacks. In order to get an authentic key chain commitment in a CDM message, a sensor only needs to receive an authentic copy of this message in most of cases since the sensor can immediately authenticate it. Though a sensor has to rely on the random selection mechanism to recover from failures, the cost is much less than those required by variations I and III. The disadvantage of Variation II is its pre-computation and storage overhead. Thus, if the base station has enough resources, Variation II should be used. Variation III sacrifices some immediate authentication capability to reduce the pre-computation and storage requirements in Variation II. Thus, if the base station has certain, but not enough resources, Variation III should be used. If the base station cannot afford the pre-computation and storage overheads required by Variation III at all, Variation I can be used to mitigate the potential DOS attacks.

2.2.7 Experimental Results

We have implemented the DOS-tolerant multi-level μTESLA scheme on TinyOS [26], which is an operating system for networked sensors. We have performed a series of experiments to evaluate the performance of the DOS-tolerant multi-level μTESLA when there are packet losses and DOS attacks against CDM messages. The communication, storage, and computation overheads are discussed in earlier sections. The focus of the evaluation in this section is on the overall effectiveness of the proposed techniques (e.g., multi-buffer random selection) in tolerating packet losses and DOS attacks, and the impact of different choices of certain parameters (e.g., buffer size, percentage of forged CDM packets). The experiments were performed using Nido, the TinyOS simulator. To simulate the lossy communication channel, we have each sensor drop each received packet with a given probability.

To further study the performance of the scheme in presence of attacks, we also implemented an attacker component, which listens to the CDM messages broadcasted by the base station and inserts forged CDM messages into the broadcast channel to disrupt the broadcast authentication. We assume that the attacker is intelligent in that it uses every piece of authentic information that a sensor node can determine in the forged messages. That is, it only modifies $K_{i+2,0}$ and the MAC value in a CDM message since any other modification can be detected by a sensor node immediately. There are other attacks against the scheme. Since they are either defeatable by the scheme (e.g., modification of data packets) or not specific to our extension (e.g., DOS attacks against the data packets), we did not consider them in our experiments.

To concentrate on the design decisions we made in our schemes, we fix the following parameters in all the experiments. We only performed the experiments with DOS-tolerant two-level μTESLA since the only purpose of having multiple levels is to scale up to a long period of time. We assume the duration of each low-level time interval is 100 ms, and each low-level key chain consists of 600 keys. Thus, the duration of each time interval for the high-level key chain is 60 seconds. We put 200 keys in the high-level key chain, which covers up to 200 minutes in time. We also set the data packet rate at base station to 100 data packets per minute. Our analysis and experiments indicate that the number of high-level keys does not have an obvious impact on the performance measures. Nevertheless, the lifetime of the two-level key chains can be extended by having more keys in the high-level key chain or another higher level of key chain. Since our purpose is to study the performance of the scheme w.r.t. to packet losses and DOS attacks, we did not do so in our evaluation.

The performance of our techniques depends on the probability of having an authentic key chain commitment, which is mainly affected by the number of CDM buffers in sensors and the percentage of forged CDM packets in the communication channel as we discussed before. Thus, in our experiments, we simply fix the CDM packet rate but use different attack rates to evaluate the performance of our system.

The performance of our system is evaluated with the following metrics: average percentage of authenticated data packets (i.e., $\frac{\#authenticated\ data\ packets}{\#received\ data\ packets}$ averaged over the sensor nodes) and average data packet authentication delay (i.e., the average time between the receipt and the authentication of a data packet). In these experiments, we focused on the impact of the following parameters on these performance metrics: sensor node's buffer size for data and CDM messages, percentage of forged CDM packets and the packet loss rate.

Because of the extremely limited memory available on sensor nodes, the buffer allocation for data packets and CDM messages becomes a major concern when we deploy a real sensor network. We evaluate the performance of different memory allocation schemes with a memory constraint. The format of data packet in our proposed technique is the same as in the original μTESLA, except for a level number, which only occupies one byte. In our implementation, both CDM and data packets consist of 29 bytes. The data packet includes a level number (1 bytes), an index (4 bytes), data (8 bytes), MAC (8 bytes) and a disclosed key (8 bytes). A CDM packet includes a level number (1 byte), an index (4 bytes), a key chain commitment $K_{i+2,0}$ (8 bytes), a MAC (8 bytes), and a disclosed key (8 bytes).

It is true that our schemes (and μTESLA) have relatively high overhead in data packets with the above settings. This is in some sense because of the small packet size. However, broadcast authentication is usually used to broadcast commands or control data from the base station to sensors. We expect typical commands or control data can fit in the 8 bytes payload. The base station also has the option to split long commands or data into multiple packets. Moreover, it is possible to modify the maximum packet size in TinyOS to decrease the overhead. In our experiment, we only consider the default maximum packet size supported by TinyOS, because the effect of CDM packets is our main concern.

When a sensor node receives a data packet, it does not need to buffer the level number and the disclosed key for future authentication; only the other 20 bytes need to be stored. For CDM packets, all copies of the same CDM message have the same values for the fields other than the key chain commitment and the MAC value (i.e., $K_{i+2,0}$ and MAC in CDM_i) since all forged messages without these values can be filtered out by the weak authentication mechanism. As a result, for all copies of CDM_i, the only fields that need saving are $K_{i+2,0}$ (8 bytes) and MAC (8 bytes), assuming that the level number and the index are used to locate the buffer and the disclosed key K_{i-1} is stored elsewhere to authenticate later disclosed keys. Further assume the totally available memory for data and CDM messages is C bytes, and the sensor node decides to store up to x data packets. Then the sensor can save up to $y = \lfloor \frac{C-20 \times x}{16} \rfloor$ copies of CDM messages.

Figure 2.8 shows the performance of different memory allocation schemes under severe DOS attacks against CDM messages (95% forged CDM packets). In these experiments, we have total memory of 512 bytes or 1K bytes. As shown in Figure 2.8, three data buffers (60 bytes) are enough to authenticate over 90% of the received data packets when the total memory is 1K bytes. This is because the data packet arrived in the later time interval carries the key that can be used to authenticate the data packets arrived in earlier time intervals. If there are no DOS attacks on data packets (such attacks are not considered in our experiments), the sensor can authenticate those data packets that arrived no less than d time intervals earlier and remove them from the buffer. Thus, the buffer size for data packets depends on the data rate, the key disclosure lag d and the duration of the lowest key chain time interval. In practice, it only needs to be large enough to hold all data packets within one lowest-level time interval.

The figure also shows that after a certain point, having more data buffers does not increase the performance. Instead, it decreases the performance since less memory is left for buffering the CDM messages.

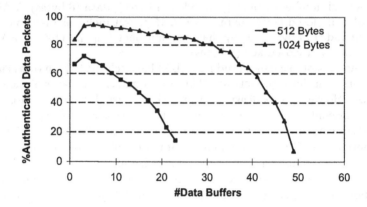

Fig. 2.8. The performance with different buffer allocation schemes for total memory 512 and 1024 bytes to buffer data and CDM messages. Assume 95% of CDM packets are forged and 50% of packets are lost when transmitted over the channel.

To measure the performance under intensive DOS attacks, we assume that each sensor node can store up to 3 data packets and 39 CDM packets, which totally occupy 684 bytes memory space. The experimental results are shown in Figures 2.9(a) and 2.9(b). Figure 2.9(a) shows that our system can tolerate DOS attacks to a certain degree; however, when there are extremely severe DOS attacks (over 95% of forged CDM packets), the performance decreases dramatically. This result is reasonable; a sensor node is certainly not able to get an authentic CDM message if all of the CDM messages it receives are forged. Nevertheless, an attacker has to make sure he/she sends many more forged CDM packets than the authentic ones to increase his/her chance of success.

Figure 2.9(a) also shows that if the base station rebroadcasts a sufficient number of CDM messages so that on average, at least one copy of such authentic CDM message can reach a sensor node in the corresponding high-level time interval (e.g., when loss rate $\leq 70\%$), the channel loss rate does not affect our scheme much. When the loss rate is large (e.g., 90% as in Figure 2.9(a)), we can observe the drop of data packet authentication rate when the percentage of forged CDM packets is low.

An interesting result is that when the channel loss rate is 90%, the data packet authentication rate initially increases when the percentage of forged CDM packets increases. This is because the sensor nodes can get the disclosed key from forged CDM packets when they cannot get it from the authentic ones.

The channel loss rate does affect the average authentication delay, which can be seen in Figure 2.9(b). The reason is that a sensor node needs to wait a longer time to get the disclosed key. Though the number of dropped packets

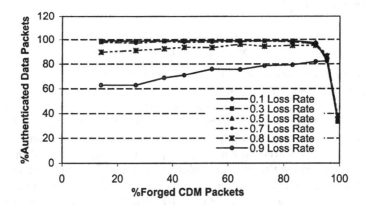

(a) Percentage of authenticated data packets

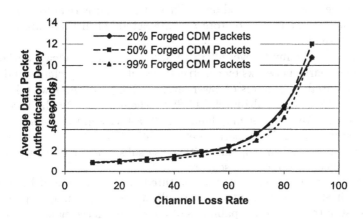

(b) Average data packet authentication delay

Fig. 2.9. Experimental results under different channel loss rate and percentage of forged CDM packets. Assuming 3 data packet buffers, 39 CDM buffers and fixed data rate (100 *data packets/minute*).

increases dramatically under severe DOS attack (over 95%) as seen in Figure 2.9(a), Figure 2.9(b) shows that the percentage of forged CDM messages does not have a significant impact on the average data packet authentication delay for those packets that have been authenticated.

In summary, the experimental results demonstrate that our system can maintain reasonable performance even with high channel loss rate under severe DOS attacks.

2.3 Tree-Based μTESLA

Despite the recent advances on broadcast authentication, several issues are still not properly addressed. Some of them are listed below.

- *Scalability (in terms of the number of senders)*: In addition to base stations, many sensor network applications may have a large number of other potential senders (e.g., mobile sinks, soldiers). For example, in battlefield surveillance where a sensor network is deployed to monitor the activities in an area, there may be a large number of soldiers or tanks, and each of them entering this area may broadcast queries to collect critical information from the network. Existing solutions require that each sensor node store the initial μTESLA parameters (e.g., the key chain commitments) for all possible senders. Although the EEPROM in sensor nodes is much more plentiful than the RAM, accessing EEPROM is much more expensive than accessing RAM. In addition, sensor nodes usually use EEPROM to store a large amount of sensed data. Thus, it is not practical for resource-constrained sensor nodes to store all parameters when there are a large number of senders.
- *DOS attacks*: The multi-level μTESLA schemes scale broadcast authentication up to large networks by constructing multi-level key chains and distributing initial parameters of lower-level μTESLA instances with higher-level ones. However, multi-level μTESLA schemes magnify the threat of DOS attacks. An attacker may launch DOS attacks on the messages carrying the initial μTESLA parameters [41, 45]. Though several solutions have been proposed in [45], they either use substantial bandwidth or require significant resources at senders.
- *Revocation* : Some senders may be captured and compromised by adversaries in hostile environments. As a result, an adversary may exploit the broadcast authentication capabilities of the compromised nodes to attack the network (e.g., consume sensors' battery power by instructing them to do unnecessary operations). Thus, it is necessary to revoke the broadcast authentication capabilities of the compromised senders once they are detected.

In this section, we develop a series of techniques to support a large number of senders and to revoke broadcast authentication capabilities from compromised senders. The proposed techniques use the μTESLA broadcast authentication protocol [63] as a building block. In other words, these techniques use multiple μTESLA instances with different parameters to provide additional capabilities related to broadcast authentication.

We assume there is an off-line *central server* with computation power and storage equivalent to a regular PC. This server is used to pre-compute and store certain parameters. We assume that the central server is well-protected. We assume that a sender has enough storage to save, or enough computation power to generate, one or several μTESLA key chains. We also assume that the clocks on sensor nodes are *loosely* synchronized, as required by the μTESLA protocol [63].

2.3.1 The Basic Approach

The essential problem in scaling up μTESLA is how to distribute and authenticate the parameters of μTESLA instances, including the key chain commitments, starting time, duration of each time interval, etc. The previous approaches construct a multi-level structure and use higher-level μTESLA instances to authenticate the parameters of lower-level ones [41, 45], thus inheriting the authentication delay introduced by μTESLA during the distribution of μTESLA parameters. The consequence of such authentication delay is that an attacker can launch DOS attacks. Moreover, they cannot handle a large number of senders. In the following, we propose to authenticate and distribute these μTESLA parameters using a Merkle hash tree [50]. This method removes the authentication delay as well as the vulnerability to DOS attacks during the distribution of μTESLA parameters and at the same time allows a large number of senders.

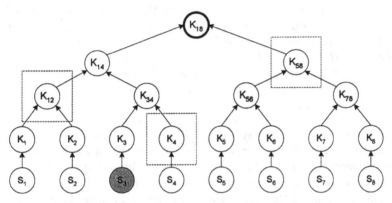

Fig. 2.10. Example of a parameter distribution tree

Assume a sensor network application requires m μTESLA instances, which may be used by different senders during different periods of time. For convenience, assume $m = 2^k$, where k is an integer. Before deployment, the central server pre-computes m μTESLA instances, each of which is assigned a unique, integer-valued ID between 1 and m. For the sake of presentation, denote the parameters (i.e., the key chain commitment, starting time, duration of each μTESLA interval, etc.) of the i-th μTESLA instance as S_i. Suppose the central server has a hash function H. The central server then computes $K_i = H(S_i)$ for all $i \in \{1, ..., m\}$, and constructs a Merkle tree [50] using $\{K_1, ..., K_m\}$ as leaf nodes. Specifically, $K_1, ..., K_m$ are arranged as leaf nodes of a full binary tree, and each non-leaf node is computed by applying H to the concatenation of its two children nodes. We refer to such a Merkle tree as a *parameter distribution tree* of parameters $\{S_1, ..., S_m\}$. Figure 2.10 shows a parameter distribution tree for eight μTESLA instances, where $K_1 = H(S_1)$, $K_{12} = H(K_1\|K_2)$, $K_{14} = H(K_{12}\|K_{34})$, etc.

The central server also constructs a *parameter certificate* for each μTESLA instance. The certificate for the i-th μTESLA instance consists of the set S_i of parameters and the values corresponding to the siblings of the nodes on the path from the i-th leaf node to the root in the parameter distribution tree. For example, the parameter certificate for the 3rd μTESLA instance in Figure 2.10 is $ParaCert_3 = \{S_3, K_4, K_{12}, K_{58}\}$. For each sender that will use a given μTESLA instance, the central server distributes the μTESLA key chain (or equivalently, the random number used to generate the key chain) and the corresponding parameter certificate to the node. The central server also predistributes the root of the parameter distribution tree (e.g., K_{18} in Figure 2.10) to regular sensor nodes, which are potentially receivers of broadcast messages.

When a sender needs to setup an authenticated broadcast channel using the i-th μTESLA instance (during a predetermined period of time), it broadcasts a message containing the parameter certificate $ParaCert_i$. Each receiver can immediately authenticate it with the pre-distributed root of the parameter distribution tree. For example, if $ParaCert_3 = \{S_3, K_4, K_{12}, K_{58}\}$ is used, a receiver can immediately authenticate it by verifying whether $H(H(K_{12}\|H(H(S_3)\|K_4))\|K_{58})$ equals the pre-distributed root value K_{18}. As a result, all the receivers can get the authenticated parameters of this μTESLA instance, and the sender may use it for broadcast authentication.

Security: According to the analysis in [61, 63], an attacker is not able to forge any message from any sender without compromising the sender itself. However, the attacker may launch DOS attacks against the distribution of parameters for μTESLA instances. Fortunately, the parameter certificates in our technique can be authenticated immediately and are immune to the DOS attacks. When a few senders are compromised, additional techniques are required to remove these compromised senders. This will be addressed in Section 2.3.4.

Overhead: In this approach, each sensor node (as a receiver) only needs to store one hash value and remember the parameters for those senders that it may communicate with. This is particularly helpful for those applications where a node only needs to communicate with a few senders or there are only a few senders staying in the network at one time.

Each sender needs to store a parameter certificate, the key chain, and other parameters (e.g., starting time) for each instance it has. To establish an authenticated broadcast channel with nodes using an instance j, a sender only needs to broadcast the corresponding pre-distributed parameter certificate, which consists of $\lceil \log m \rceil$ hash values and the parameter set S_j. This is practical since such distribution only needs to be done once for each instance.

After receiving this parameter certificate, a sensor node only needs $1 + \lceil \log m \rceil$ hash functions to verify the related parameters.

Comparison: Compared with the multi-level μTESLA schemes [41, 45], the most significant gain of the proposed approach is the removal of the authentication delay in distributing the μTESLA parameters. The multi-level μTESLA schemes are subject to DOS attacks against the distribution of μTESLA parameters because of the authentication delay. Specifically, receivers cannot authenticate parameter distribution messages immediately after receiving them, and thus have to buffer such messages. An attacker may send a large amount of bogus messages to consume receivers' buffers and thus launch DOS attacks. To mitigate or defeat such DOS attacks, the multi-level μTESLA schemes either use duplicated copies of distribution messages along with a multi-buffer, random selection strategy, or require substantial pre-computation at the sender. In contrast, the proposed approach does not have these problems. With the proposed approach, senders may still duplicate parameter distribution messages to deal with communication failures. However, unlike multi-level μTESLA schemes, a sender does not have to compete with malicious attackers in terms of the number of messages with this approach. In other words, with the proposed approach it is sufficient for a receiver to get one copy of each parameter distribution message.

In general, our approach allows late binding of μTESLA instances with senders. For example, the central server may reserve some μTESLA instances during deployment time and distribute them to mobile sinks as needed during the operation of the sensor networks. This allows us to add new senders dynamically by simply generating enough number of instances at the central server for later joined senders. Thus, in our later discussion, we will not discuss how to add new senders.

There are multiple ways to arrange senders, μTESLA instances, and their parameters in a parameter distribution tree. Different ways may have different properties. Next we investigate one specific scheme that has some additional attractive properties. We consider other options as future work.

2.3.2 A Scheme for Long-Lived Senders

The following discussion presents a special instantiation of the basic approach when there are up to m senders in the network and n_j μTESLA instances for each sender j. The purpose is to improve the parameter distribution for those senders that may stay in the network for a long period of time. The protocol can be divided into two phases: *pre-distribution* and *establishment of authenticated broadcast channel*.

Pre-Distribution: The central server first divides the (long) lifetime of each sender into n_j time intervals such that the duration of each time interval (e.g., 1 hour) is suitable for running a μTESLA instance on a sender and sensor nodes efficiently. For convenience, we denote such a time interval as a *(μTESLA) instance interval* , or simply an *instance interval*. When $n_j = 1$

for all $j \in \{1, ..., m\}$, the long-lived version becomes the basic scheme. (Note that each instance interval should be partitioned into smaller time intervals (e.g., 500ms intervals) to run the μTESLA protocol.)

For sender j, the central server generates one μTESLA instance for each instance interval. The corresponding key chains are linked together by pseudo random functions. Specifically, the central server generates the last key of the n_j-th μTESLA key chain randomly; for the i-th μTESLA key chain, the central server generates the last key by performing a pseudo random function F' on the first key (the key next to the commitment) of the $(i+1)$-th μTESLA key chain. Let $S_{j,i}$ denote the parameters (e.g., key chain commitment, starting time) of the i-th μTESLA instance for sender j. The parameters (such as the duration of each μTESLA interval) that can be pre-determined do not need to be included in $S_{j,i}$.

For each sender j, the central server generates a parameter distribution tree $Tree_j$ from $\{S_{j,1}, \cdots, S_{j,n_j}\}$. This tree is used to distribute the parameters of different μTESLA instances for sender j. Let $ParaCert_{j,i}$ denote the parameter certificate for $S_{j,i}$ in $Tree_j$. Assume R_j is the root of $Tree_j$. The central server then generates the parameter distribution tree $Tree_R$ for all senders from $\{S_1, \cdots, S_m\}$, where S_j consists of R_j and parameters that are not included in $\{S_{j,1}, \cdots, S_{j,n_j}\}$ for sender j. If a parameter (e.g., n_j) is the same for all senders, it can be pre-distributed to all sensor nodes before the deployment of sensor networks. Let $ParaCert_j$ denote the parameter certificate for S_j in $Tree_R$. The central server pre-distributes to each sender j the parameter certificate $ParaCert_j$, the n_j μTESLA instances, and the parameter distribution tree $Tree_j$. The central server also pre-distributes the root value of tree $Tree_R$ to each sensor node. Figure 2.11 shows an example of the above construction.

Establishment of Authenticated Broadcast Channel: To establish an authenticated broadcast channel with sensor nodes, a sender j first broadcasts $ParaCert_j$ to authenticate the root S_j. The authenticity of S_j can be verified as discussed in the basic approach. To distribute the parameters of the i-th μTESLA instance, sender j only needs to $ParaCert_{j,i}$, which can be verified by re-computing the root R_j from $ParaCert_{j,i}$ if S_j, which includes R_j, is already verified. After this, the sender j can authenticate the messages using the i-th μTESLA instance, and the sensor nodes having the authentic parameters can verify the broadcast messages from sender j. To deal with message loss, a sender may broadcast $ParaCert_j$ and $ParaCert_{j,i}$ multiple times.

Security: In the long-lived scheme, different key chains are linked together. This does not sacrifice security. The knowledge of the commitment of later key chain cannot be used to recover the first key in the later key chain and thus cannot be used to recover any key in earlier key chains since it is computationally infeasible to revert the pseudo random function. Moreover, this instantiation is resistant to DOS attacks if each parameter certificate can be delivered in one packet, similar to the basic approach. When it is necessary

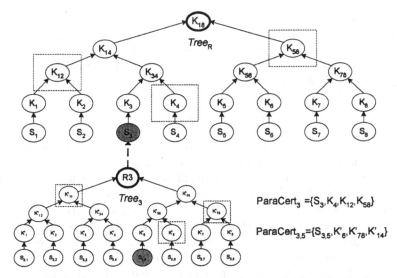

Fig. 2.11. Example of a parameter distribution tree for long-lived schemes

to send a parameter certificate in multiple packets, there may be DOS attacks. We will investigate this problem in Section 2.3.3.

Overhead: The above scheme requires each sender j to store n_j key chains, a parameter certificate $ParaCert_j$, and a parameter distribution tree $Tree_j$. This storage overhead is usually affordable at senders since they may be much more resourceful than the sensor nodes. Similar to the basic scheme, each sensor node only needs to store one hash value, the root of $Tree_R$. To establish an authenticated broadcast channel with sensor nodes, sender j needs to broadcast $ParaCert_j$, which includes $\lceil \log m \rceil$ hash values and parameters in S_j, and $ParaCert_{j,i}$, which includes $\lceil \log n_j \rceil$ hash values and parameters in $S_{j,i}$. A sensor node needs to perform $1 + \lceil \log m \rceil$ hash functions to verify the root R_j of tree $Tree_j$ and $1 + \lceil \log n_j \rceil$ hash functions to verify the corresponding parameters.

Comparison: Compared with the basic approach, this scheme has several benefits. First, the parameters of the i-th μTESLA instance for each sender j is divided into the distribution of the parameters common to all μTESLA instances (i.e., S_j) for the same sender and those specific to each μTESLA instance (i.e., $S_{j,i}$). Thus, the communication overhead can be reduced. Second, this scheme connects different μTESLA key chains together through pseudo random functions, and thus provides two options to verify a disclosed μTESLA key. A sensor node can always verify disclosed keys with an earlier key. This is suitable when there are no long term communication failures or channel jamming attacks since a sensor node can usually authenticate a disclosed key with a few pseudo random functions. Alternatively, a sensor node can authenticate any disclosed key using the commitment derived from the most recently

verified parameter certificate. This is suitable when there are long term communication failures or channel jamming attacks, or for newly deployed nodes.
Third, when all μTESLA instances are linked together, a sensor node may use
a later key to derive an earlier key to authenticate a buffered message. Such
a capability is not available for independent μTESLA instances.

2.3.3 Distributing Parameter Certificates

As we mentioned earlier, the proposed technique is resistant to the DOS attacks if each parameter certificate is delivered in one packet since a receiver
can authenticate such a certificate immediately upon receiving it. However,
due to the low bandwidth and small packet size in sensor networks, a certificate may be too large to be transmitted in a single packet. As a result, it is
often necessary to fragment each certificate and deliver it in multiple packets.

A straightforward approach is to simply split those values in a certificate
into multiple packets. However, this simple idea suffers from DOS attacks,
where an attacker sends a large number of forged certificates and forces a
sensor node to perform a lot of computations to identify the right one from
those fragments. To deal with this problem, we adopt the idea in [32]. Intuitively, we fragment a parameter certificate in such a way that a sensor
node can authenticate each fragment independently instead of trying every
combination.

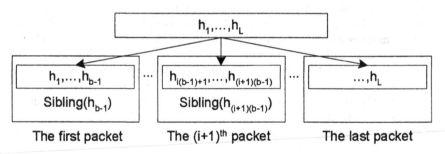

Fig. 2.12. Example of fragmentation

Assume a parameter certificate then consists of L values $\{h_1, h_2, \cdots, h_L\}$,
and each packet can carry b values. As shown in Figure 2.12, in the first step
of fragmentation, we put the first $b-1$ values in the first packet, the second
$b-1$ values in the second packet, and so on, until there are no more values left.
If the last packet only includes one value, we move it to the previous packet
and remove the last packet. The previous packet then becomes the last packet,
containing b values. In the second step, we append in every packet other than
the last one the sibling (in the parameter distribution tree) of the last value in
this packet. By doing this, the first fragment can be authenticated immediately
once the sensor node receives an authentic fragment. After authenticating the

first fragment, the second fragment can be also authenticated immediately using the values in the first fragment. This process will continue until the sensor node receives all authentic fragments.

For example, as shown in Figure 2.10, $ParaCert_3$ consists of 4 values, $\{K_{58}, K_{12}, K_4, S_3\}$. Assume that each fragment can carry 3 hash values and S_3 consists of 1 key chain commitment. Using the above technique, the first packet includes $\{K_{58}, K_{12}, K_{34}\}$, and the second packet includes K_4, S_3. If a sensor node receives the first fragment, it can authenticate the fragment by verifying whether $H(H(K_{12}|K_{34})|K_{58})$ equals the pre-distributed root value. Once the first fragment is authenticated successfully, the second fragment can be authenticated by verifying if $H(H(S_3)|K_4)$ equals the hash value K_{34}, which is contained in the first fragment.

The above technique can significantly reduce the computation overhead required to authenticate the certificate fragments when there are DOS attacks. In addition, if most of fragments are received in order, there is no need to allocate a large buffer to store these fragments since most of fragments can be authenticated immediately.

2.3.4 Revoking μTESLA Instances

In hostile environments, not only sensor nodes but also broadcast senders may be captured and compromised by adversaries. Once a sender is compromised, the attacker can forge any broadcast message using the secrets stored on this sender and convince other sensor nodes to perform unnecessary or malicious operations. Thus, it is necessary to revoke the broadcast authentication capability from compromised senders.

We do not consider the process or techniques to detect compromised or captured broadcast senders here, but assume such results are given. The detection of compromised or captured senders is in general difficult, but feasible at least in certain scenarios. For example, in battlefields, broadcast senders may be carried by soldiers or unmanned vehicles. If a soldier or an unmanned vehicle is captured, we need to revoke its broadcast authentication capability.

We propose two approaches to revoke compromised senders. The first one uses a revocation tree to take back the broadcast authentication capability from compromised senders; the second one employs proactive refreshment to control the broadcast authentication capability of each sender. Revocation of compromised senders requires the central server to be on-line when it broadcasts revocation messages; however, the central server can still remain off-line in other situations.

Revocation Tree: When a sender is detected to have been compromised, the central server broadcasts a revocation message with the IDs of the sender. This message has to be authenticated; otherwise, an attacker may forge such messages to revoke non-compromised senders. We may use another μTESLA instance maintained by the central server to authenticate such messages. However, this instance has special functions and may become an attractive target for DOS attacks due to the authentication delay. The following discussion provides an alternative method that does not suffer from DOS attacks or authentication delay.

The main idea of this method is to construct a Merkle tree similar to parameter distribution trees, which is called a *revocation tree*, since its purpose is to revoke broadcast authentication capabilities from compromised senders. The revocation tree is built from sender IDs and random numbers. If the sender ID j and the corresponding random number is disclosed in an authenticated way, sender j is revoked.

Assume there are potentially m senders. For simplicity, we assume $m = 2^k$ for an integer k. The central server generates a random number r_j for each sender with ID j, where $1 \leq j \leq m$. The central server then constructs a Merkle tree where the j-th leaf node is the concatenation of ID j and r_j. We refer to this Merkle tree as the *revocation tree*. The central server finally distributes the root of the revocation tree to all sensor nodes. We assume the central server is physically secure. Protection of the central server is an important but separate issue, which is not addressed here.

When a sender j is detected to have been compromised, the central server broadcasts the ID j and the random number r_j. To authenticate these values, the central server has to broadcast the sibling of each node on the path from "$j||r_j$" (i.e., the leaf node for j in the revocation tree) to the root. This is exactly the same as the parameter certificate technique used to authenticate μTESLA parameters. To distinguish from parameter certificate, we refer to the above set of values as a *revocation certificate*, denoted $RevoCert_j$. With $RevoCert_j$, any sensor node can recompute the root hash value and verify it by checking whether it leads to the pre-distributed root value. If a sensor node gets a positive result from this verification, it puts the corresponding sender into a revocation list and stops accepting broadcast messages from the sender. To deal with message loss, the distribution of a revocation certificate may be repeated multiple times.

The revocation tree approach cannot guarantee the revocation of all compromised senders in presence of communication failures, though traditional fault tolerant techniques can provide high confidence. However, it guarantees that a non-compromised sender will not be revoked. This is because the revocation of a sender requires a revocation certificate, which is only known to the central server. An attacker cannot forge any revocation certificate without access to the random numbers kept in the leaves of the revocation tree, due to the one-way function used to generated the revocation tree [50].

In this approach, each sensor node needs to store an additional hash value, the root of the revocation tree. To revoke a sender, the central server distributes a revocation certificate, which consists of $1 + \lceil \log m \rceil$ values. An overly long revocation certificate can be transmitted in the same way as discussed in the previous subsection. To authenticate the revocation certificate, a sensor node needs to perform $1 + \lceil \log m \rceil$ hash functions.

The revocation tree approach has several limitations. First, due to the unreliable wireless communication and possible malicious attacks (e.g., channel jamming), the revocation messages are not guaranteed to reach every sensor node. As a result, an attacker can convince those sensor nodes that missed the revocation messages to do unnecessary or malicious operations using the revoked μTESLA instances. Second, each sensor node needs to store a revocation list, which introduces additional storage overhead, especially when a large number of senders are revoked.

Proactive Refreshment of Authentication Keys: To deal with the limitations of the revocation tree approach, we present an alternative method to revoke the authentication capability from compromised senders. The basic idea is to distribute a fraction of authentication keys to each sender and have the central server update the keys for each sender when necessary. A clear benefit is that if a sender is compromised, the central server only needs to stop distributing new authentication keys to this sender; there is no need to broadcast a revocation message and maintain a revocation list at each sensor node. In addition, this approach guarantees that once compromised senders are detected, they will be revoked from the network after a certain period of time. The authentication keys for each sender can be distributed in a proactive way since we can predetermine the time when a key will be used.

Specifically, during the pre-distribution phase, the central server distributes the parameter certificates (but not the μTESLA instances) to each sender. For simplicity, we assume that the central server gives a μTESLA instance to a sender each time. Before the current μTESLA instance expires, the central server distributes the key used to derive the next μTESLA key chain to the sender through a key distribution message encrypted with a key shared between the central server and the sender if the sender has not been detected to have been compromised. The sender may then generate the next μTESLA key chain accordingly. To increase the probability of successful distribution of authentication keys in presence of communication failures, the central server may send each key distribution message multiple times.

As mentioned earlier, the revocation of a compromised sender is guaranteed (with certain delay) in the proactive refreshment approach when it is detected to have been compromised. However, the broadcast authentication capability of a sender is not guaranteed if there are message losses. A sender may miss all key distribution messages that carry new authentication keys due to unreliable wireless communication and malicious attacks. Thus, a sender may have no keys to authenticate new data packets. Moreover, there may be a long delay between the detection and the revocation of a compromised sender, and the compromised sender may still have keys that can be used to forge broadcast messages.

In the proactive refreshment approach, instead of storing n_j μTESLA instances, a sender j only needs to store a few of them. Thus, the storage overhead is reduced. However, the communication overhead between the central server and the senders is increased since the central server has to distribute keys to each sender individually. Moreover, the central server has to be on-line more often. There are no additional communication and computation overheads for sensor nodes.

Both of the above approaches have advantages and disadvantages. In practice, these two options may be combined to provide better performance and security. The revocation certificates from the central server can mitigate the problem of the delay between the detection and the revocation of a compromised sender, while the proactive refreshment technique guarantees the future revocation of a compromised sender if the compromise is detected.

2.3.5 Implementation and Evaluation

We have implemented the long-lived version of the proposed techniques on TinyOS [26] and used Nido , the TinyOS simulator, to evaluate the performance. Our evaluation is focused on the broadcast of data packets and the distribution of μTESLA parameters. Since the number of senders does not affect these two aspects, we only consider a single sender in the evaluation.

We compare our techniques with the multi-level μTESLA schemes in [41, 45], where two schemes were proposed: multi-level DOS-tolerant μTESLA and multi-level DOS-resistant μTESLA. Multi-level DOS-resistant μTESLA has as much communication overhead as multi-level DOS-tolerant μTESLA and will fall back to multi-level DOS-tolerant μTESLA at a receiver when the receiver misses all copies of a parameter distribution message. Thus, we only compare our scheme with the multi-level DOS-tolerant μTESLA, which can be obtained from http://discovery.csc.ncsu.edu/software/ML-microTESLA.

We adopt a setting similar to [41, 45]: the μTESLA key disclosure delay is 2 μTESLA time intervals, the duration of each μTESLA time interval is 100 ms, and each μTESLA key chain consists of 600 keys. Thus, the duration of each μTESLA instance is 60 seconds. We assume that there are 200 μTESLA instances, which cover up to 200 minutes in time. Each parameter set $S_{j,i}$ only contains a μTESLA key chain commitment. This means that each parameter certificate contains 9 hash values. Assume each hash value, cryptographic key or MAC value is 8 bytes long. The parameter certificate can be delivered with 4 packets, each of which contains a sender ID (2 bytes), a key chain index (2 bytes), a fragment index (1 byte), and 3 hash values (24 bytes). As a result, the packet payload size is 29 bytes, which is the default maximum payload size in TinyOS [26].

The multi-level μTESLA schemes use Commitment Distribution Messages (CDM) to distribute parameters of μTESLA instances. According to the implementation we obtained, each CDM message in multi-level DOS-tolerant scheme also contains 29 bytes payload. For convenience, we call a parameter certificate fragment or a CDM packet a *parameter distribution packet*.

We study and compare the performances of the tree-based technique and the multi-level DOS-tolerant μTESLA scheme in terms of DOS attacks, channel loss rate, and storage and communication overheads. We set the data packet rate from the sender as 100 data packets per minute, and allocate 3 buffers for data packets at each sensor node. The metrics we are interested in here are the *authentication rate* , which is the fraction of authenticated data packets; the *distribution rate* , which is the fraction of successfully distributed parameters; and the *average failure recovery delay* , which is the average number of μTESLA time intervals needed to have the authenticated parameters for the next μTESLA instance after a sensor node loses every authentic parameter distribution message for a given μTESLA instance.

We use a simple strategy to rebroadcast a parameter certificate, where the rebroadcasts of any certificate are non-interleaving, and the fragments of each certificate are broadcast in order. Other strategies are also possible. However, we consider them as possible future work.

To investigate the authentication rate and the distribution rate under DOS attacks and communication failures, we assume the attacker sends 200 forged parameter distribution packets per minute. We also assume the channel loss rate is 0.2. Figure 2.13(a) illustrates the authentication rate for both schemes as the frequency of parameter distribution packets increases. We assume 20 CDM buffers at each receiver for the multi-level DOS-tolerant μTESLA scheme. We can see that the tree-based scheme always has a higher authentication rate than the multi-level DOS-tolerant μTESLA scheme. The reason is that in the tree-based scheme, a sensor node is able to authenticate any buffered message once it receives a later disclosed key since different key chains are linked together. Although in the multi-level DOS-tolerant μTESLA scheme, lower-level μTESLA key chains are also linked to the higher-level ones, a sensor node may have to wait for a long time to recover an authentication key from the higher-level key chain when the corresponding lower-level key chain commitment is lost due to severe DOS attacks or channel losses. During this time period, most of the previous buffered data packets are already dropped.

Figure 2.13(b) shows the authentication rate for both schemes as the number of buffers for parameter distribution packets increases. We assume that the sender distributes 20 parameter distribution packets per minute. We can see that the multi-level DOS-tolerant μTESLA scheme has to allocate a large buffer to achieve a certain authentication rate when there are severe DOS attacks, while the tree-based scheme can achieve a higher authentication rate without any additional buffer. The reason is that in the tree-based scheme, a sensor node can verify a parameter certificate immediately and thus there is no need to buffer certificates; while in the multi-level DOS-tolerant μTESLA scheme, a sensor node has to wait for a while before authenticating CDM messages.

Figure 2.14(a) focuses on the communication and storage overhead introduced by both schemes to achieve high distribution rate. It shows that to achieve a desirable distribution rate under severe DOS attacks, the multi-level DOS-tolerant μTESLA scheme requires a large buffer and a high rebroadcast

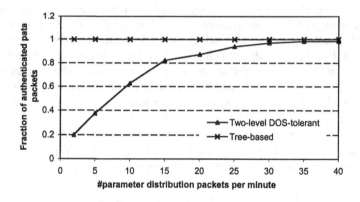

(a) 20 CDM buffers for multi-level μTESLA

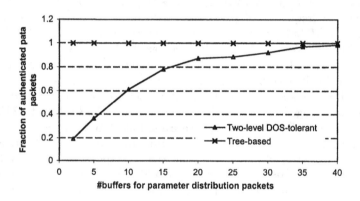

(b) 20 parameter distribution packets per minute

Fig. 2.13. Authentication rate under 0.2 loss rate and 200 forged parameter distribution packet per minute.

frequency, while the tree-based scheme is more communication and storage efficient. Note that the number of μTESLA key chains supported by the tree-based scheme affects the number of fragments for each certificate and thus affects the distribution rate. Figure 2.14(b) shows that to achieve high distribution rate, the number of key chains supported by the tree-based scheme increases dramatically when the frequency of parameter distribution packets increases. This means that the tree-based scheme can cover a very long time period by increasing a little communication overhead.

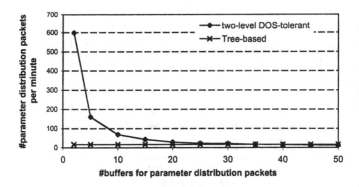

(a) Communication overhead v.s. storage overhead

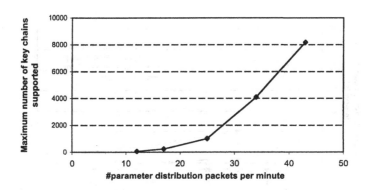

(b) Maximum number of key chains v.s. frequency of distributing parameter distribution packets

Fig. 2.14. Channel loss rate: 0.2; # forged commitment distribution: 200 per minute; distribution rate: 95%.

To investigate the average failure recovery delay, we assume that the sender distributes 20 parameter distribution packets per minute. (Note that the multi-level DOS-resistant μTESLA scheme has to fall back to the multi-level DOS-tolerant μTESLA scheme if a sensor node loses every authentic copy of a given CDM message.)

Figure 2.15(a) shows the average failure recovery delay for both schemes as the channel loss rate increases. We assume 20 CDM buffers for the multi-level μTESLA scheme. We can see that the average failure recovery delay of the tree-based scheme increases with the channel loss rate, while the multi-

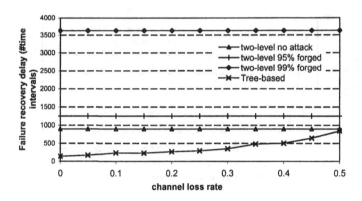

(a) 20 CDM buffers for multi-level μTESLA.

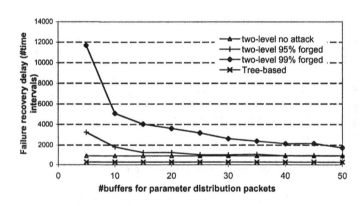

(b) 0.2 channel loss rate

Fig. 2.15. Average failure recovery delay. Assume 20 parameter distribution packet per minute.

level μTESLA scheme is not affected when the loss rate is small. However, the recovery delay of the multi-level μTESLA scheme increases rapidly when there are severe DOS attacks. In contrast, the tree-based scheme is not affected by DOS attacks if the attacker does not jam the channel completely. Since the channel loss rate is usually a small value, the tree-based scheme has shorter recovery delay than the multi-level μTESLA scheme in most cases. Figure 2.15(b) shows the impact of storage overhead on the average failure recovery delay. We assume that the channel loss rate is 0.2. The average failure recovery delay of the multi-level μTESLA scheme increase quickly when the number of buffers for parameter distribution packets decreases, while the tree-based scheme has shorter delay and is not affected by the number of buffers for parameter distribution packets.

2.4 Summary

In this chapter, we first developed a multi-level key chain scheme to efficiently distribute the key chain commitments for the broadcast authentication scheme named μTESLA. By using pre-determination and broadcast, our approach removed μTESLA's requirement of a unicast-based distribution of initial key chain commitments, which introduces high communication overhead in large distributed sensor networks. We also proposed several techniques, including periodic broadcast of commitment distribution messages and random selection strategies, to improve the survivability of our scheme and defeat some DOS attacks. The resulting protocol, named multi-level μTESLA, satisfies several nice properties, including low overhead, tolerance of message loss, scalability to large networks, and resistance to replay attacks as well as DOS attacks.

We then identified a number of new challenges in broadcast authentication for wireless sensor networks. Several practical tree-based broadcast authentication techniques were developed to support multiple senders, distribute parameters for μTESLA instances, and revoke the broadcast authentication capabilities of compromised senders in wireless sensor networks. Our analysis and experiment show that the tree-based techniques are efficient and practical and have better performance than previous approaches.

Note that all the proposed schemes require loosely time synchronization, which may not be true in some applications. There are many mechanisms to disrupt the time synchronization method. Thus, it is particularly desirable to have alternative ways of authenticating broadcast messages without the assumption of time synchronization.

3

Pairwise Key Establishment

In this chapter, we develop a number of key pre-distribution techniques to deal with the pairwise key establishment in sensor networks. After reviewing some existing techniques for pairwise key establishment in sensor networks, we present a general framework for pairwise key establishment based on the polynomial-based key pre-distribution scheme [6] and the probabilistic key pre-distribution scheme [20, 12]. This framework is called the *polynomial pool-based* key pre-distribution scheme , which uses a polynomial pool instead of a key pool in [20, 12]. The secrets on each sensor node are generated from a subset of polynomials in the pool. If two sensor nodes have the secrets generated from the same polynomial, they can establish a pairwise key based on the polynomial-based key pre-distribution scheme. All the previous schemes in [6, 20, 12] can be considered as special instances in this framework.

By instantiating the components in this framework, we further develop two novel pairwise key pre-distribution schemes: a *random subset assignment* scheme and a *hypercube-based* scheme. The random subset assignment scheme assigns each sensor node the secrets generated from a random subset of polynomials in the polynomial pool. The hypercube-based scheme arranges polynomials in a hypercube space, assigns each sensor node to a unique coordinate in the space, and gives the node the secrets generated from the polynomials related to the corresponding coordinate. Based on this hypercube, each sensor node can then identify whether it can directly establish a pairwise key with another node, and if not, what intermediate nodes it can contact to indirectly establish the pairwise key.

Our analysis indicates that our new schemes have some nice features compared with the previous methods. In particular, when the fraction of compromised secure links is less than 60%, given the same storage constraint, the random subset assignment scheme provides a significantly higher probability of establishing secure communication between non-compromised nodes than the previous methods. Moreover, unless the number of compromised nodes sharing a common polynomial exceeds a threshold, compromise of sensor nodes does not lead to the disclosure of keys established between non-compromised nodes using this polynomial.

Similarly, the hypercube-based scheme also has a number of attractive properties. First, it guarantees that any two nodes can establish a pairwise key when there are no compromised nodes, provided that the sensor nodes can communicate with each other. Second, it is resilient to node compromise. Even if some sensor nodes are compromised, there is still a high probability to re-establish a pairwise key between non-compromised nodes. Third, a sensor node can directly determine whether it can establish a pairwise key with another node and how to compute the pairwise key if it can. As a result, there is no communication overhead during the discovery of directly shared keys.

To reduce the computation at sensor nodes, we provide an optimization technique for polynomial evaluation. The basic idea is to compute multiple pieces of key fragments over some special finite fields such as F_{2^8+1} and $F_{2^{16}+1}$, and concatenate these fragments into a regular key. A nice property provided by such finite fields is that no division is necessary for modular multiplication. As a result, evaluation of polynomials can be performed efficiently on low-cost processors on sensor nodes that do not have division instructions. Our analysis indicates that such a method only slightly decreases the uncertainty of the keys.

We have implemented these algorithms on MICA2 motes [14] running TinyOS [26]. The implementation only occupies a small amount of memory (e.g. 416 bytes in ROM and 20 bytes in RAM for one of our implementations, excluding the memory for polynomial coefficients). The evaluation indicates that computing a 64-bit key using this technique can be faster than generating a 64-bit MAC (Message Authentication Code) using RC5 [67] or SkipJack [58] for a reasonable degree of polynomial. These results show that our schemes are practical for resource-constrained sensor networks.

3.1 Key Pre-Distribution Techniques in Sensor Networks

This section reviews two types of techniques to perform key pre-distribution in the context of resource constrained sensor networks: the polynomial-based key pre-distribution [6] and the probabilistic key pre-distribution [20, 12, 64].

3.1.1 Polynomial-Based Key Pre-Distribution

The original key pre-distribution protocol in [6] was developed for group key pre-distribution. Since the goal is to establish pairwise keys, for simplicity, we only discuss the special case of pairwise key establishment in the context of sensor networks.

To pre-distribute pairwise keys, the (key) setup server randomly generates a bivariate t-degree polynomial $f(x,y) = \sum_{i,j=0}^{t} a_{ij} x^i y^j$ over a finite field F_q, where q is a prime number that is large enough to accommodate a cryptographic key, such that it has the property of $f(x, y) = f(y, x)$. (In the following, we assume that all the bivariate polynomials have this property without explicit statement.) It is assumed that each sensor node has a unique ID. For each node i, the setup server computes a *polynomial share* of $f(x, y)$, that is, $f(i, y)$. This polynomial share is pre-distributed to node i. Thus, for any two sensor nodes i and j, node i can compute the key $f(i, j)$ by evaluating $f(i, y)$ at point j. And node j can compute the same key $f(j, i) = f(i, j)$ by evaluating $f(j, y)$ at point i. As a result, nodes i and j can establish a common key $f(i, j)$.

In this approach, each sensor node i needs to store a t-degree polynomial $f(i, x)$, which occupies $(t + 1) \log q$ storage space. To establish a pairwise key, both sensor nodes need to evaluate the polynomial at the ID of the other sensor node. There is no communication overhead during the pairwise key establishment process.

The security proof in [6] ensures that this scheme is unconditionally secure and t-collusion resistant . That is, the coalition of no more than t compromised sensor nodes knows nothing about the pairwise key between any two non-compromised nodes.

It is theoretically possible to use the general group key distribution protocol in [6] in sensor networks. However, the storage cost for a polynomial share is exponential in terms of the group size, making it prohibitive in sensor networks. This chapter focuses on the problem of pairwise key establishment.

3.1.2 Probabilistic Key Pre-Distribution

A probabilistic key pre-distribution technique was proposed to bootstrap initial trust between sensor nodes in [20]. The main idea is to have each node randomly pick a set of keys from a key pool before deployment so that any two sensor nodes can share a common key with certain probability. Specifically, a setup server, which is assumed to be trusted, generates a large pool of random keys, where each key has a unique key ID. Each sensor node then gets assigned a random subset of keys as well as their IDs from this pool before the deployment of this sensor node.

In order to establish a common key directly between two sensor nodes after deployment, the nodes only need to identify a common key ID they share. This can be achieved by exchanging the list of key IDs they have. The probability of sharing at least one common key can be easily derived for a given key pool size s and the number of keys s' at sensor nodes using the following equation [20].

$$p = 1 - \prod_{i=0}^{s'-1} \frac{s - s' - i}{s - i}$$

For example, if the key pool size is 100,000 and each sensor node randomly gets assigned 200 keys, the probability of sharing at least one common key is

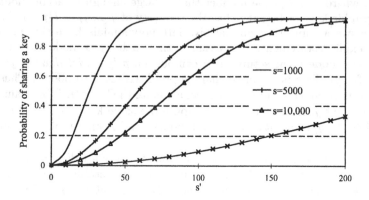

Fig. 3.1. Probability of sharing at least one key for different combinations of key pool size and the number of keys at sensor nodes.

about 0.33. Figure 3.1 shows the probability of sharing at least one key for different combinations of key pool size and the number of keys assigned to each sensor node.

Note that it is possible that two sensor nodes cannot establish a common key directly. In this case, they need to find a number of other sensor nodes to help them establish a temporary session key. A simple way is to find another node that can directly establish keys with both the source and the destination nodes and let this node act as an intermediate node (like a KDC) between them.

Chan et al. extended the basic probabilistic key pre-distribution scheme and proposed the q-composite key pre-distribution scheme [12]. This approach requires two sensor nodes to setup a pairwise key only when they share at least q common keys. This idea improves the security of the basic probabilistic scheme [20] when there are a small number of compromised sensor nodes. Pietro et al. proposed a seed-based key deployment strategy to simplify the key discovery procedure and a cooperative protocol to enhance its performance [64].

The advantage of the above key pre-distribution schemes is that they require a small amount of memory at sensor nodes but guarantee a high probability of sharing a common key between two sensor nodes. The main disadvantage is that a small number of compromised sensor nodes discloses a large fraction of secrets in the network, which is shown in Figure 3.2. The reason is that the same key may be shared by more than two sensor nodes.

Chan et al. also developed a random pairwise keys scheme to defeat node capture attacks [12]. In this scheme, the setup server randomly selects a pair of sensor nodes and distributes a unique random key between them. The benefit

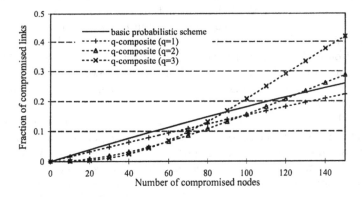

Fig. 3.2. Fraction of compromise links between non-compromised sensor nodes for the basic probabilistic scheme and the q-composite scheme when $p = 0.33$ and $s' = 200$.

is that none of the keys shared directly between non-compromised sensor nodes will be disclosed no matter how many sensor nodes are compromised. The disadvantage is that the maximum supported network size is limited by the storage overhead and the desired probability of sharing a key between sensor nodes directly.

Du et al. independently discovered a technique similar to one of our proposed polynomial pool-based key pre-distribution schemes (random subset assignment) [18]. However, our scheme provides a more general framework, which makes it possible to discover novel key pre-distribution schemes. The idea of using prior deployment knowledge was independently discovered in [17] and our work [43] to improve the key pre-distribution schemes.

3.2 Polynomial Pool-Based Key Pre-Distribution

The polynomial-based key pre-distribution scheme discussed in Section 3.1.1 has some limitations. In particular, it can only tolerate the collusion of no more than t compromised nodes, where the value of t is limited by the available memory space and the computation capability on sensor nodes. Indeed, the larger a sensor network is, the more likely an adversary compromises more than t sensor nodes and then the entire network.

To have secure and practical key establishment techniques, we develop a general framework for key pre-distribution based on the scheme presented in Section 3.1.1. We call it the *polynomial pool-based key pre-distribution* since a pool of random bivariate polynomials are used in this framework. In this section, we focus on the discussion of this general framework. In the next two sections, we will present two efficient instantiations of this framework.

The polynomial pool-based key pre-distribution is inspired by the studies in [20, 12]. The basic idea can be considered as the combination of the polynomial-based key pre-distribution and the key pool idea used in [20, 12]. However, our framework is more general in that it allows different choices to be instantiated within this framework, including those in [20, 12] and our later instantiations in sections 3.3 and 3.4.

Intuitively, this general framework generates a pool of random bivariate polynomials and assigns shares on a subset of bivariate polynomials in the pool to each sensor node. The polynomial pool has two special cases. When it has only one polynomial, the general framework degenerates into the polynomial-based key pre-distribution. When all the polynomials are 0-degree, the polynomial pool degenerates into a key pool used in [20, 12].

Pairwise key establishment in this framework has three phases: *setup, direct key establishment* , and *path key establishment* . The setup phase is performed to initialize the nodes by distributing polynomial shares to them. After being deployed, if two sensor nodes need to establish a pairwise key, they first attempt to do so through direct key establishment. If they can successfully establish a common key, there is no need to start path key establishment; otherwise, these two nodes start path key establishment, trying to establish a pairwise key with the help of other sensor nodes.

3.2.1 Phase 1: Setup

The setup server randomly generates a set $\mathcal{F}$ of bivariate t-degree polynomials over the finite field F_q. To identify different polynomials, the setup server may assign each polynomial a unique ID. For each sensor node i, the setup server picks a subset of polynomials $\mathcal{F}_i \subseteq \mathcal{F}$ and assigns the shares of these polynomials to node i. The main issue in this phase is the *subset assignment* problem , which specifies how to pick a subset of polynomials from $\mathcal{F}$ for each sensor node.

Here we identify two ways to perform subset assignments: *random assignment* and *pre-determined assignment* .

Random Assignment

With random assignment, the setup server randomly picks a subset of $\mathcal{F}$ for each sensor node. This random selection should be evenly distributed in $\mathcal{F}$ for security concerns; otherwise, some polynomials may have higher probability of being selected and higher frequency of being used in key establishment than the others and thus become the primary targets of attacks. Several parameters may be used to control this process, including the number of polynomial shares assigned to a node and the size of $\mathcal{F}$. In the simplest case, the setup server assigns the same number of random selected polynomial shares to each sensor node.

Pre-Determined Assignment

When pre-determined assignment is used, the setup server follows a certain scheme to assign subsets of $\mathcal{F}$ to sensor nodes. A pre-determined assignment should bring some nice properties that can be used to improve direct and path key establishment.

3.2.2 Phase 2: Direct Key Establishment

A sensor node starts Phase 2 if it needs to establish a pairwise key with another node. If both sensor nodes have shares on the same bivariate polynomial, they can establish the pairwise key directly using the polynomial-based key pre-distribution discussed in Section 3.1.1. The main issue in this phase is the *polynomial share discovery* problem, which specifies how to find a common bivariate polynomial of which both nodes have polynomial shares. For convenience, we say that two sensor nodes have a *secure link* if they can establish a pairwise key through direct key establishment. A pairwise key established in this phase is called a *direct key* .

Here we identify two types of techniques to solve this problem: *pre-distribution* and *real-time discovery*.

Pre-Distribution

The setup server pre-distributes certain information to the sensor nodes so that given the ID of another sensor node, a sensor node can determine whether it can establish a direct key with the other node. A naive method is to let each sensor node store the IDs of all the sensor nodes with which it can setup direct keys. However, this naive method has difficulties in dealing with the sensor nodes that join the network on the fly, because the setup server has to inform some existing nodes about the addition of new sensor nodes. Alternatively, the setup server may map the ID of each sensor node to the IDs of polynomial shares it has so that given the ID of a sensor node, anybody can derive the IDs of polynomial shares it has. Thus, any sensor node can determine immediately whether it can establish a direct key with a given sensor node by only knowing its ID. Note that this method requires the pre-determined assignment strategy in the setup phase.

The drawback of pre-distribution methods is that an attacker may also know the distribution of the polynomial shares. As a result, the attacker may precisely target certain sensor nodes, attempting to learn the shares of a particular bivariate polynomial. The following alternative way may avoid this problem.

Real-time Discovery

Intuitively, real-time discovery requires two sensor nodes to discover on the fly whether they have shares on a common bivariate polynomial. As one possible way, two nodes may first exchange the IDs of polynomials of which they both have shares and then try to identify the common polynomial. To protect the IDs of the polynomials, the sensor node may challenge the other party to solve puzzles instead of disclosing the IDs of the polynomials directly. Similar to the the method in [20], when node i needs to establish a pairwise key with node j, it sends node j an encryption list, $\alpha, E_{K_v}(\alpha), v = 1, ..., |\mathcal{F}_i|$, where K_v is computed by evaluating the vth polynomial share in $\mathcal{F}_i$ on point j (a potential pairwise key node j may have). When node j receives this encryption list, it first computes $\{K'_v\}_{v=1,...,|\mathcal{F}_j|}$, where K'_v is computed by evaluating the vth polynomial share in $\mathcal{F}_j$ on point i (a potential pairwise key node i may have). Node j then generates another encryption list $\{E_{K'_v}(\alpha)\}_{v=1,...,|\mathcal{F}_j|}$. If there exists a common encryption value that is included in both encryption lists, node i and node j can establish a common key, which is the key used to generate this common value.

The drawback of real-time discovery is that it introduces additional communication overhead that does not appear in the pre-distribution approaches. If the polynomial IDs are exchanged in clear text, an attacker may gradually learn the distribution of polynomials among sensor nodes and selectively capture and compromise sensor nodes based on this information. However, it is more difficult for an adversary to collect the polynomial distribution information in the real-time discovery method than in the pre-distribution method since the adversary has to monitor the communication among sensor nodes. In addition, when the encryption list is used to protect the IDs of polynomial shares in a sensor node, an adversary has no way to learn the polynomial distribution among sensor nodes and thus cannot launch selective node capture attacks.

3.2.3 Phase 3: Path Key Establishment

If direct key establishment fails, two sensor nodes need to start Phase 3 to establish a pairwise key with the help of other sensor nodes. To establish a pairwise key with node j, a sensor node i needs to find a sequence of nodes between itself and node j such that any two adjacent nodes in this sequence can establish a direct key. For the sake of presentation, we call such a sequence of nodes a *key path* (or simply a *path*) since the purpose of such a path is to establish a pairwise key. Then either node i or j initiates a key establishment request with the other node through the intermediate nodes along the path. A pairwise key established in this phase is called an *indirect key* . A subtle issue is that two adjacent nodes in the path may not be able to communicate with each other directly. In this work, we assume that they can always discover a route between themselves so that the messages from one node can reach the other.

The main issue in this phase is the *path discovery* problem , which specifies how to find a path between two sensor nodes. Similar to Phase 2, there are two types of techniques to address this problem.

Pre-Distribution

Similar to the pre-distribution technique in phase 2, the setup server pre-distributes certain information to each sensor node so that given the ID of another node, each node can determine at least one key path to the other node directly. The resulting key path is called the *pre-determined key path.* For convenience, we call the process to compute the pre-determined key paths *key path pre-determination* . The drawback is that an attacker may also take advantage of the pre-distributed information to attack the network. Moreover, it is possible that none of the pre-determined key paths is available to establish an indirect key between two nodes due to compromised (intermediate) nodes or communication failures.

To deal with the above problem, the source node needs to dynamically find other key paths to establish an indirect key with the destination node. For convenience, we call such a process *dynamic key path discovery* . For example, the source node may contact a number of other nodes with which it can establish direct keys using non-compromised polynomials, attempting to find a node that has a path to the destination node to help establish an indirect key.

Real-time discovery

Real-time discovery techniques have the sensor nodes discover key path on the fly without any pre-determined information. The sensor nodes may take advantage of the direct keys established through direct key establishment. For example, to discover a key path to another sensor node, a sensor node picks a set of intermediate nodes with which it has established direct keys. The source node may send requests to all these intermediate nodes. If one of the intermediate nodes can establish a direct key with the destination node, a key path is discovered. Otherwise, this process may continue with the intermediate nodes forwarding the request. Such a process is similar to a route discovery process used to establish a route between two nodes. The drawback is that such methods may introduce substantial communication overhead.

3.3 Key Pre-Distribution Using Random Subset Assignment

In this section, we present the first instantiation of the general framework by using a random strategy for the subset assignment during the setup phase. That is, for each sensor node, the setup server selects a random subset of polynomials in $\mathcal{F}$ and assigns the corresponding polynomial shares to the node.

3.3.1 The Random Subset Assignment Scheme

The random subset assignment scheme can be considered an extension to the basic probabilistic scheme in [20]. Instead of randomly selecting keys from a large key pool and assigning them to sensor nodes, our method randomly chooses polynomials from a polynomial pool and assigns their polynomial shares to sensor nodes. However, our scheme also differs from the scheme in [20]. In [20], the same key may be shared by multiple sensor nodes. In contrast, in our scheme, there is a unique key for each pair of sensor nodes. If no more than t shares on the same polynomial are disclosed, none of the pairwise keys constructed using this polynomial between two non-compromised sensor nodes will be disclosed.

Now let us describe this scheme by instantiating the three components in the general framework.

1. **Subset assignment:** The setup server randomly generates a set $\mathcal{F}$ of s bivariate t-degree polynomials over the finite field F_q. For each sensor node, the setup server randomly picks a subset of s' polynomials from $\mathcal{F}$ and assigns shares as well as the IDs of these s' polynomials to the sensor node.

2. **Polynomial share discovery:** Since the setup server does not pre-distribute enough information to the sensor nodes for polynomial share discovery, sensor nodes that need to establish a pairwise key have to find a common polynomial with real-time discovery techniques. To discover a common bivariate polynomial, the source node discloses a list of polynomial IDs to the destination node. If the destination node finds that they have shares on the same polynomial, it informs the source node the ID of this polynomial; otherwise, it replies with a message that contains a list of its polynomial IDs, which also indicates that the direct key establishment fails.

3. **Path discovery:** If two sensor nodes fail to establish a direct key, they need to start the path key establishment phase. During this phase, the source node tries to find another node that can help it setup a pairwise key with the destination node. Basically, the source node broadcasts two list of polynomial IDs. One includes the polynomial IDs at the source node, and the other includes the polynomial IDs at the destination node. These two lists are available at both the source and the destination nodes after the polynomial share discovery. If one of the nodes that receives this request is able to establish direct keys with both the source and the destination nodes, it replies with a message that contains two encrypted copies of a randomly generated key: one encrypted by the direct key with the source node, and the other encrypted by the direct key with the destination node. Both the source and the destination nodes can then get the new pairwise key from this message. (Note that the intermediate node acts as an ad hoc KDC in this case.) In practice, we may restrict that a sensor node only contact its neighbors within a certain range.

3.3.2 Performance

Similar to the analysis in [20], the probability of two sensor nodes sharing the same bivariate polynomial, which is the probability that two sensor nodes can establish a direct key, can be estimated by

$$p = 1 - \prod_{i=0}^{s'-1} \frac{s - s' - i}{s - i}. \tag{3.1}$$

Figure 3.3(a) shows the relationship between p and the combinations of s and s'. It is easy to see that the closer s and s' are, the more likely two sensor nodes can establish a direct key. Our later security analysis (in Section 3.3.4) shows that small s and s' can provide high security performance. This differs from the the basic probabilistic scheme [20] and the q-composite scheme [12], where the key pool size has to be very large to meet certain security requirement. The reason is that there is another parameter (i.e., the degree t of the polynomials) that affects the security performance of the random subset assignment scheme. In Equation 3.1, the value of s' is affected by the storage overhead and the degree t of the polynomials. In fact, we have $t = \frac{C}{s'} - 1$, where C is the number keys a sensor node can store.

Now let us consider the probability that two nodes can establish a key through either the polynomial share discovery or the path discovery. Let d denote the average number of neighbor nodes that each sensor node contacts. Consider any one of these d nodes. The probability that it shares direct keys with both the source and the destination nodes is p^2, where p is computed by Equation 3.1. As long as one of these d nodes can act as an intermediate node, the source and the destination nodes can establish a common key. It follows that the probability of two nodes establishing a pairwise key (directly or indirectly) is $P_s = 1 - (1 - p)(1 - p^2)^d$. Figure 3.3(b) shows that the probability P_s of establishing a pairwise key between two sensor nodes increases quickly as the probability p of establishing direct keys or the number d of neighbor nodes it contacts increases.

3.3.3 Overheads

Each node has to store s' t-degree polynomials over F_q, which introduces $s'(t+1) \log q$ bits storage overhead. In addition, each node needs to remember the IDs of revoked nodes with which it can establish direct keys. Assume the IDs of sensor nodes are chosen from a finite field $F_{q'}$. The storage overhead introduced by the revoked IDs is at most $s't \log q'$ bits since if $t+1$ shares of one bivariate polynomial are revoked, this polynomial is compromised and can be discarded. Thus, the overall storage overhead is at most $s'(t+1) \log q + s't \log q'$ bits.

In terms of communication overhead, during the polynomial share discovery, the source node needs to disclose a list of s' IDs to the destination node.

The communication overhead is mainly due to the transmission of such lists. During the path discovery, the source node broadcasts a request message that consists of two lists of polynomial IDs. This introduces one broadcast message at the source node and possibly several broadcast messages at other nodes receiving this request if they further forward this request. However, due to the small values of s and s' in our scheme, all the broadcast messages are small and can be replayed efficiently in resource-constrained sensor networks. If a node receiving this message shares common polynomials with both the source and the destination nodes, it only needs to reply with a message consisting of two encrypted copies of a randomly generated key.

In terms of computational overhead, the polynomial share discovery requires one polynomial evaluation at each node if two nodes share a common polynomial. During the path discovery, if a node receiving the request shares common polynomials with both the source and the destination nodes, it only needs to perform two polynomial evaluations and two encryptions. If there exists at least one intermediate node that can be used in the establishment of an indirect key, both the source and the destination nodes only need one decryption.

3.3.4 Security Analysis

According to the security analysis in [6], an attacker cannot determine keys established with a particular polynomial if he/she has compromised no more than t sensor nodes that have shares of this polynomial. Assume an attacker randomly compromises N_c sensor nodes, where $N_c > t$. Consider any polynomial f in $\mathcal{F}$. The probability of f being chosen for a sensor node is $\frac{s'}{s}$, and the probability of this polynomial being chosen exactly i times among N_c compromised sensor nodes is

$$P[i \text{ compromised shares}] = \frac{N_c!}{(N_c - i)!i!}(\frac{s'}{s})^i(1 - \frac{s'}{s})^{N_c-i}.$$

Thus, the probability of a particular bivariate polynomial being compromised can be estimated as $P_{cd} = 1 - \sum_{i=0}^{t} P[i \text{ compromised shares}]$. Since f is any polynomial in $\mathcal{F}$, the fraction of compromised links between non-compromised nodes can be estimated as P_{cd}. Figure 3.4(a) includes the relationship between the fraction of compromised links for non-compromised nodes and the number of compromised nodes for some combinations of s and s'. We can see that the random subset scheme provides high security guarantee in terms of the fraction of compromised links between non-compromised nodes when the number of compromise nodes does not exceed certain threshold. (To save space, Figure 3.4 also includes the performance of the basic probabilistic scheme [20] and the q-composite scheme [12], which will be used for the comparison in Section 3.3.5.)

If an attacker knows the distribution of polynomials over sensor nodes, he/she may target specific sensor nodes in order to compromise the keys de-

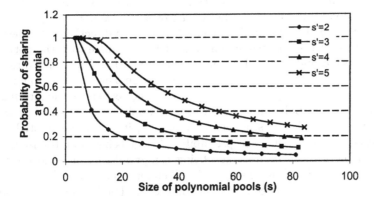

(a) The probability p that two nodes share a polynomial v.s. the size s of the polynomial pool

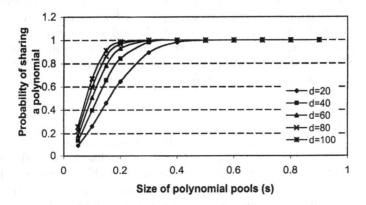

(b) The probability P_s of establishing a pairwise key v.s. the probability p that two nodes share a polynomial

Fig. 3.3. Probabilities about pairwise key establishment

rived from a particular polynomial. In this case, the attacker only needs to compromise $t + 1$ sensor nodes. However, targeting specific nodes is generally more difficult than randomly compromising sensor nodes since the attacker has to compromise the *selected* nodes.

An easy fix to remove the above threat is to restrict that each polynomial be used for at most $t + 1$ times. As a result, an attacker cannot recover a polynomial unless he/she compromises all related sensor nodes. Though effective at improving the security, this method also puts a limit on the maximum number of sensor nodes for a given combination of s and s'. Indeed, given the above constraint, the total number of sensor nodes cannot exceed $\frac{(t+1)\cdot s}{s'}$.

To estimate the probability of any (direct or indirect) key between two non-compromised nodes being compromised, we assume that the network is fully connected and each pair of nodes can establish a direct or indirect key. Thus, among all pairwise keys, there are a fraction p of direct keys and a fraction $1-p$ of indirect keys on average. For each indirect key, if the intermediate node and the two polynomials used in the establishment of this key are not compromised, the key is still secure; otherwise, it cannot be trusted. Moreover, each direct key has the probability P_{cd} of being compromised. Thus, the probability of an indirect key being compromised can be estimated by $1-(1-p_c)(1-P_{cd})^2$, where $p_c = \frac{N_c}{N}$. Therefore, the probability of any (direct or indirect) key between two non-compromised nodes being compromised can be estimated by

$$P_c = p \times P_{cd} + (1-p)[1-(1-p_c)(1-P_{cd})^2].$$

Figure 3.4(b) includes the relationship between the fraction of compromised (direct or indirect) keys for non-compromised nodes and the number of compromised nodes for some combinations of s and s'. We can see that the random subset scheme also provides high security guarantee in terms of the fraction of compromised (direct or indirect) keys between non-compromised nodes when the number of compromise nodes does not exceed a certain threshold.

Two non-compromised sensor nodes may need to re-establish an indirect key when the current pairwise key is compromised. The knowledge of the identities of compromised nodes is generally a difficult problem, which needs deep investigation. However, when such detection mechanism is available and the node compromises are detected, it is always desirable to re-establish the pairwise key. Thus, we assume that the detection of compromised nodes is done through other techniques, and is not considered in this work.

Assume the source node contacts d neighbor nodes to re-establish an indirect key with the destination node. Among these d nodes, the average number of non-compromised nodes can be estimated by $\frac{d(N-N_c)}{N}$ for simplicity. If one of these non-compromised nodes shares common non-compromised polynomials with both the source and the destination nodes, a new pairwise key can be established. Thus, the probability of re-establishing an indirect key between two non-compromised nodes can be estimated by

$$P_{re} = 1 - [1 - p^2(1-P_{cd})^2]^{\frac{d(N-N_c)}{N}}.$$

Figure 3.5 includes the relationship between the probability to re-setup an indirect key for non-compromised nodes and the number of compromised nodes in the network. It shows that there is still a high probability to re-establish a pairwise key between two non-compromised nodes when the current key is compromised, given that the network still provides certain security performance (e.g., less than 60% compromised links).

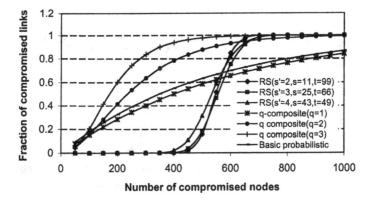

(a) Fraction of compromised links between non-compromised nodes v.s. number of compromised nodes

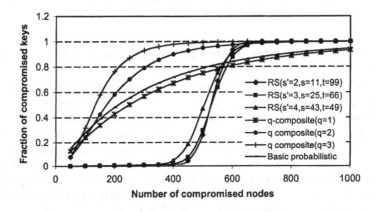

(b) Fraction of compromised keys (direct or indirect) between non-compromised nodes v.s. number of compromised nodes. Assume $N = 20,000$

Fig. 3.4. Performance of the random subset assignment scheme under attacks. RS refers to the random subset assignment scheme. Assume each node has available storage for 200 keys and $p = 0.33$.

3.3.5 Comparison with Previous Schemes

The random subset assignment scheme has a number of advantages compared with the basic probabilistic scheme [20], the q-composite scheme [12], and the random pairwise keys scheme [12]. In this analysis, we first compare the communication and computational overheads introduced by these schemes

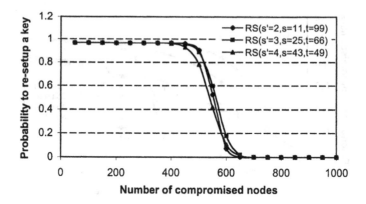

Fig. 3.5. The probability of re-establishing a pairwise key using path discovery. Assume each node has available storage equivalent to 200 keys, and contacts 30 neighbor nodes $d = 30$. Assume $N = 20,000$

given certain storage constraint, and then compare their security performance under attacks.

We do not compare the random subset assignment scheme with the multiple-space key pre-distribution scheme in [18] since these two schemes are actually equivalent to each other. In fact, in the multiple-space key pre-distribution scheme, the elements in the second row of matrix G can be considered as the IDs of sensor nodes in the random subset assignment scheme; each matrix D_i can be considered as the coefficients of a bivariate polynomial; each row in a matrix A_i can be considered as a polynomial share; computing a key through $A_c(i) \times G(j)$ can be considered as evaluating a polynomial share.

After the direct key establishment, the basic idea of the path key establishment is to find an intermediate node that shares direct keys with both the source and the destination nodes. This is similar in all the previous schemes and the random subset assignment scheme. For simplicity, we focus on the overheads in direct key establishment. Note that each coefficient in our scheme takes about the same amount of space as a cryptographic key since F_q is a finite field that can just accommodate the keys. We assume that each sensor node can store up to C keys or polynomial coefficients.

The communication and computational overheads for different schemes are summarized in Table 3.1. The communication overhead is calculated using the size of the list of key or polynomial IDs; the computational overhead is calculated using the number of comparisons in identifying the common key or polynomial and the number of polynomial evaluations, assuming that the IDs of keys or polynomials are stored in ascend order in each node and binary search is used to locate the ID of the common key or polynomial.

Table 3.1. Communication and computational overheads for direct key establishment in different schemes. s_k is the key pool size in the basic probabilistic scheme and the q-composite scheme. $s' = \frac{C}{t+1}$. The last row will be discussed in Section 3.4.

	Communication	Computation
Basic probabilistic scheme [20]	$C \log s_k$	$\frac{2C+p-pC}{2} \log C$ comparisons
q-composite scheme [12]	$C \log s_k$	$C \log C$ comparisons
Random pairwise keys scheme [12]	0	0
Random subset assignment scheme	$s' \log s$	$\frac{2s'+p-ps'}{2} \log s'$ comparisons + 1 polynomial evaluation
Grid-based scheme	0	1 polynomial evaluation

Comparison with the Basic Probabilistic and the q-Composite Scheme

According to Table 3.1, we can see that the random subset assignment is usually much more efficient than the basic probabilistic scheme [20] and the q-composite scheme [12] in terms of communication overhead due to small s and s'. Indeed, this overhead is reduced by a factor of at least $t + 1$. However, the computation overhead is more expensive in the random subset assignment scheme since it has to evaluate a t-degree polynomial.

Figures 3.4(a) and 3.4(b) compare the security performance of the random subset assignment scheme with the basic probabilistic scheme [20] and the q-composite scheme [12]. These figures clearly show that before the number of compromised sensor nodes reaches a certain point, the random subset assignment scheme performs much better than both of the other schemes. When the number of compromised nodes exceeds a certain point, the other schemes have fewer compromised links or keys than the random subset assignment scheme. Nevertheless, under such circumstances, none of these schemes provide sufficient security, due to the large fraction of compromised links (over 60%) or the large fraction of compromised (direct or indirect) keys (over 80%). Thus, the random subset assignment scheme clearly has advantages over the basic probabilistic scheme [20] and the q-composite scheme [12].

Comparison with the Random Pairwise Keys Scheme

As shown in Table 3.1, the random pairwise keys scheme [12] does not have any communication and computational overheads in direct key establishment since it stores the IDs of all other nodes with which it can establish direct keys.

In terms of security performance, the random pairwise keys scheme does not allow reuse of the same key by multiple pairs of sensor nodes [12]. Thus,

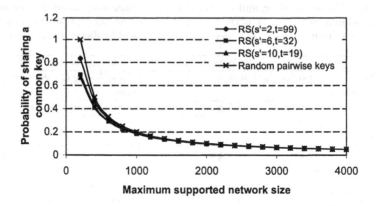

Fig. 3.6. The relationship between the probability of establishing a common key and the maximum supported network size in order to be resilient against node compromise.

the compromise of some sensor nodes does not lead to the compromise of direct keys shared between non-compromised nodes. As we discussed earlier, with a restriction that no polynomial be used more than $t + 1$ times, our scheme can ensure the same property.

Now we compare the performance between the random subset assignment scheme under the above restriction and the random pairwise keys scheme. The maximum number of nodes that the random subset assignment scheme supports can be estimated as $N = \frac{s \times (t+1)}{s'}$. Assuming the storage overhead in each sensor node is $C = s' \cdot (t+1)$, we have $s = \frac{N \times s'^2}{C}$. Together with Equation 3.1, we can derive the probability of establishing a direct key between two nodes with a given storage constraint. Figure 3.6 plots the probability of two sensor nodes sharing a direct key in terms of the maximum network size for the random pairwise keys scheme [12] and the random subset assignment scheme under restriction. We can easily see that the random subset assignment scheme under restriction has lower but almost the same performance as the random pairwise keys scheme.

The random subset assignment scheme has several advantages over the random pairwise keys scheme [12]. In particular, in the random subset assignment scheme, sensor nodes can be added dynamically without having to contact the previously deployed sensor nodes. In contrast, in the random pairwise keys scheme, if it is necessary to dynamically deploy sensor nodes, the setup server has to either reserve space for sensor nodes that may never be deployed, which reduces the probability that two deployed nodes share a common key, or inform some previously deployed nodes of additional pairwise keys, which introduces additional communication overhead. Moreover, given certain storage constraint, the random subset assignment scheme (without the restriction on the reuse of polynomials) allows the network to grow, while the random pairwise keys scheme has an upper limit on the network size. Thus, the random subset assignment scheme would be a more attractive choice than the random pairwise keys scheme in certain applications.

3.4 Hypercube-based Key Pre-Distribution

In this section, we give another instantiation of the general framework, which we call the *hypercube-based* key pre-distribution. This scheme has a number of attractive properties. First, it guarantees that any two sensor nodes can establish a pairwise key when there are no compromised sensor nodes, assuming that the nodes can communicate with each other. Second, this scheme is resilient to node compromises. Even if some nodes are compromised, there is still a high probability of re-establishing a pairwise key between two non-compromised nodes. Third, a sensor node can directly determine whether it can establish a direct key with another node, and if it can, which polynomial should be used. As a result, there is no communication overhead during polynomial share discovery.

Note that in the preliminary version of this work [42], we studied a *grid-based* key pre-distribution technique. Hypercube-based key pre-distribution can be considered as a generalization of grid-based key pre-distribution. The grid-based key pre-distribution scheme is interesting due to its simplicity. However, we do not explicitly discuss it here because of space reasons. Please refer to [42] for details.

3.4.1 The Hypercube-Based Scheme

Given a total of N sensor nodes in the network, the hypercube-based scheme constructs an n-dimensional hypercube with m^{n-1} bivariate polynomials arranged for each dimension j, $\{f^j_{\langle i_1,...,i_{n-1}\rangle}(x,y)\}_{0 \le i_1,...,i_{n-1} < m}$, where $m = \lceil \sqrt[n]{N} \rceil$. Figure 3.7(a) shows a special case of the hypercube-based scheme when $n = 2$ and $m = 5$ (i.e., the grid-based scheme). In this figure, each column i is associated with a polynomial $f^1_i(x,y)$, and each row i is associated with a polynomial $f^2_i(x,y)$. The setup server then assigns each node in the network to a unique coordinate in this n-dimensional space. For the sensor node at coordinate $(j_1,...,j_n)$, the setup server pre-distributes the polynomial shares of $\{f^1_{\langle j_2,...,j_n\rangle}(x,y),...,f^n_{\langle j_1,...,j_{n-1}\rangle}(x,y)\}$ to this node. As a result, sensor nodes can perform share discovery and path discovery using this pre-distributed information.

For convenience, we encode a node's coordinate in the hypercube into a single-valued node ID. Every valid coordinate in the hypercube is first converted into n l-bit binary strings (one from each dimension), where

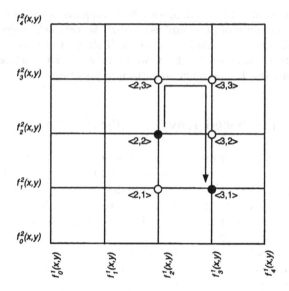

(a) An example of hypercube when $n = 2$ and $m = 5$

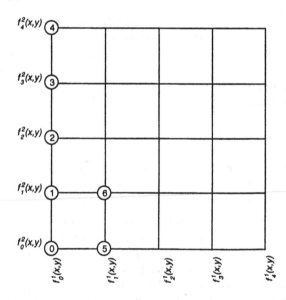

(b) An example order of node assignment

Fig. 3.7. Hypercube-based key pre-distribution when $n = 2$ and $m = 5$

$l = \lceil \log_2 m \rceil$. These n binary strings are then concatenated to generate an integer value which is used as the ID of the node. In our discussion, we conceptually represent each ID j as $\langle j_1, ..., j_n \rangle$, where j_i is called the *sub-index* of ID j in dimension i, which also represents the ith l bits of j.

1. **Subset assignment:** The setup server randomly generates $n \times m^{n-1}$ t-degree bivariate polynomials $\mathcal{F} = \{f^j_{\langle i_1, ..., i_{n-1} \rangle}(x, y) \mid 1 \leq j \leq n, 0 \leq i_1, ..., i_{n-1} < m\}$ over a finite field F_q. For each sensor node, the setup server selects an unoccupied coordinate $(j_1, ..., j_n)$ in the n-dimensional space and assigns it to this node. This coordinate $\langle j_1, ..., j_n \rangle$ is then used as the ID of this node. The setup server then distributes $\{ID,$ $f^1_{\langle j_2, ..., j_n \rangle}(j_1, y), ..., f^n_{\langle j_1, ..., j_{n-1} \rangle}(j_n, y)\}$ to this sensor node. To facilitate path discovery and guarantee that at least one key path exists when there are no compromised nodes and any two nodes can communicate with each other, we always select the coordinate corresponding to the smallest unassigned ID. Specifically, the setup server assigns the ith sensor node that requests for polynomial shares the coordinate $(a_1, a_2, \cdots, a_n)$, where $a_j = \lfloor \frac{i}{m^{n-j}} \rfloor \mod m^{n-j+1}$ for $1 \leq j \leq n$. Figure 3.7(b) shows a possible order to assign coordinates to sensor nodes when $n = 2$ and $m = 5$. It is easy to see that if there exist nodes at $\langle i, j \rangle$ and $\langle i', j' \rangle$, then there must be a node at either $\langle i, j' \rangle$ or $\langle i', j \rangle$, or both.

2. **Polynomial share discovery:** To establish a pairwise key with node j, node i checks whether they have the same sub-indexes in $n-1$ dimensions. In other words, it checks the Hamming distance d_h between their IDs i and j. If $d_h = 1$, nodes i and j share a common polynomial, and they can establish a direct key using the polynomial-based key pre-distribution scheme; otherwise, they need to go through path discovery to establish an indirect key. For example, if $j_k = i_k$ for all $1 \leq k \leq n - 1$ ($d_h = 1$), both nodes i and j have polynomial shares of $f^n_{\langle j_1, ..., j_{n-1} \rangle}(x, y)$ and thus can use this polynomial to establish a direct key.

3. **Path discovery:** If nodes i and j cannot establish a direct key, they need to find a key path between each other in the hypercube. For example, in Figure 3.7(a), both of node $\langle 2, 1 \rangle$ and $\langle 3, 2 \rangle$ can help node $\langle 2, 2 \rangle$ establish a pairwise key with node $\langle 3, 1 \rangle$. Indeed, if there are no compromised nodes and any two nodes can communicate with each other, it is guaranteed that there is at least one key path which can be used to establish a session key between any two nodes, due to the node assignment algorithm. In fact, nodes i and j can pre-determine such a key path using the following key path pre-determination algorithm without communicating with others. Assume $i > j$ if we consider node IDs i and j as integer values. The following algorithm can be performed on either of them.

   a) The source node maintains a set $\mathcal{L} = \{d_1, ..., d_w\}_{d_1 < d_2 \cdots < d_w}$ that records the dimensions that nodes i and j have different sub-indexes, a list $\mathcal{P}$ that records the key path computed by this algorithm, a most recently computed intermediate node u and the largest ID having been assigned. Initially, $\mathcal{P}$ is a list with a single node i and $u = i$.

b) Given u and $\mathcal{L}$, the next intermediate node v is computed by randomly selecting an element d' in $\mathcal{L}$ so that $v = \langle u_1, \cdots, u_{d'-1}, j_{d'}, u_{d'+1}, \cdots, u_n \rangle$ is not larger than the largest ID having been assigned (or i if this is not available). The algorithm then removes d' from $\mathcal{L}$, appends v to $\mathcal{P}$, and lets $u = v$. If $\mathcal{L}$ is empty, it appends j to $\mathcal{P}$ and returns $\mathcal{P}$ as the discovered key path; otherwise it repeats this step.

The correctness of the above key path pre-determination algorithm is guaranteed by Lemma 3.1. Once such a key path is computed, the source node i can send a request to the destination node j along the key path to establish an indirect key. For example, if $i = \langle 1, 3, 5 \rangle$ and $j = \langle 0, 2, 4 \rangle$, the key path "$\langle 1, 3, 5 \rangle$, $\langle 0, 3, 5 \rangle$, $\langle 0, 2, 5 \rangle$, $\langle 0, 2, 4 \rangle$" is one of those paths. To establish an indirect key with j, the source node i sends a key establishment request to node $\langle 0, 3, 5 \rangle$ and node $\langle 0, 3, 5 \rangle$ forwards the request to node $\langle 0, 2, 5 \rangle$, which further forwards the request to the destination node j. Every message transmitted between two adjacent nodes in the key path is encrypted and authenticated with the direct key shared between them.

Lemma 3.1. *The above key path pre-determination algorithm guarantees to compute a key path between any two sensor nodes.*

Proof. We first show the size of $\mathcal{L}$ is reduced by 1 each time Step (b) is executed. To prove this, we need to prove that given u and $\mathcal{L}$, there exists at least one element d' in $\mathcal{L}$ so that $v = \langle u_1, \cdots, u_{d'-1}, j_{d'}, u_{d'+1}, \cdots, u_n \rangle$ is not larger than the largest ID having been assigned. Note that every $u_{d'}$ is either $i_{d'}$ or $j_{d'}$. Consider d_1. Since $i > j$, we have $i_{d_1} > j_{d_1}$. Thus, if $u_{d_1} = i_{d_1}$, we choose $d' = d_1$ and compute the next node v. It is easy to verify that $v < i$. If $u_{d_1} = j_{d_1}$ (d_1 has been chosen before), we have $u_{d_1} < i_{d_1}$. This implies that $v < i$ for any d' in $\mathcal{L}$. Thus, we can choose any value in $\mathcal{L}$. As a result, u can always find the next node v, and the size of the set $\mathcal{L}$ is reduced by 1 each time Step (b) is executed. Eventually, the above key pre-determination algorithm will output a sequence of nodes with node j as the last node. Moreover, the Hamming distance between u and v in the second step is exactly 1. This implies that every two adjacent nodes in $\mathcal{P}$ can establish a direct key. Thus, we can conclude that the above key path pre-determination algorithm guarantees to compute a key path between any two sensor nodes.

3.4.2 Dynamic Key Path Discovery

Though the path discovery algorithm described above can pre-determine a key path with a number of intermediate nodes, the intermediate nodes may have been compromised, or are out of communication range in some situations. However, there are alternative key paths. In particular, we may reuse the pre-determined paths at other nodes to find a secure key path. For example, in Figure 3.7(a), besides node $\langle 2, 1 \rangle$ and $\langle 3, 2 \rangle$, node $\langle 2, 3 \rangle$ has a pre-determined path to node $\langle 3,1 \rangle$ through node $\langle 3,3 \rangle$. Thus, it can help node i setup a common key with node j.

Though it is possible to flood the network to find a key path, the resource constraints on sensor nodes make this method impractical. Instead, we propose the following algorithm to find a key path between nodes S and D dynamically. The basic idea is to have the source node and each intermediate node contact a non-compromised node that is "closer" to the destination node in terms of the Hamming distance between their IDs. Indeed, if there are no compromised nodes in the network, the above key path pre-determination algorithm can always find a key path if any two nodes can communicate with each other. In practice, we may use the dynamic path discovery instead to achieve better performance when there are attacks or communication failures. To increase the chance of success, the following algorithm may be performed multiple rounds. It is assumed that every message between two nodes in the algorithm is encrypted and authenticated with the direct key between them.

1. In order to establish an indirect key with node D, node S generates a random number r and maintains a counter c with initial value 0. In each round, it increments c and computes $K_c = F(r, c)$, where F is a pseudo random function [23]. Then, it constructs a message $M = \{S, D, K_c, c, flag\}$ with $flag = 1$ and goes to the next step.

   (The $flag$ in message M indicates whether the Hamming distance is reduced by forwarding M to the next intermediate node. The purpose is to control the length of the path discovered by this algorithm and the number of messages.)

2. Consider a sensor node u having the message $M = \{S, D, K_c, c, flag\}$. Node u first tries to find a non-compromised node v that can establish a direct key with u using a non-compromised polynomial and has a smaller Hamming distance to D than u. If this succeeds, u sets $flag$ in M to 1 and sends the modified message M to v. We can see that the Hamming distance between v and D is one smaller than that between u and D.

   If u cannot find such a node and $flag$ in M is 0, the path discovery stops. Otherwise, it selects a non-compromised node v that can establish a direct key with u using a non-compromised polynomial and has the same Hamming distance to D as u. If u finds such a node v, it sets $flag$ in message M to 0 and sends to v the modified message M. If it cannot find such a node, the path discovery protocol at this node stops.

3. When the destination node D receives the key establishment request, it knows that node S wants to setup a pairwise key with it. Node D then sets the pairwise key as $K_{S,D} = K_c$ and informs node S the counter value c. As a result, these sensor nodes share the same pairwise key.

Lemma 3.2. *For any two nodes S and D, the above dynamic key path discovery algorithm guarantees to find a key path with $d_h - 1$ intermediate nodes if there are no compromised nodes and any two nodes can communicate with each other, where d_h is the Hamming distance between S and D.*

Proof. Let $\mathcal{L} = \{d_1, ..., d_w\}_{d_1 < d_2 \cdots < d_w}$ records the dimensions in which nodes u and D have different sub-indexes. If $\mathcal{L}$ only has one element, S and D can establish a direct key and the path discovery succeeds.

Now assume $\mathcal{L}$ contains more than one element. We show that any intermediate node u can find a dimension e so that u and D have different sub-indexes in this dimension and the node $\langle u_1, ..., u_{e-1}, D_e, u_{e+1}, ..., u_n \rangle$ exists. Consider dimension d_1. If $u > D$, we have $u_{d_1} > D_{d_1}$. Thus, u can choose node $v = \langle u_1, ..., u_{d_1-1}, D_{d_1}, u_{d_1+1}, ..., u_n \rangle$. Since $v < u$, v must be the ID of an existing node. If $u < D$, we can choose any value in $\mathcal{L}$ other than d_1 since we always have $v < D$ and v must be the ID of an existing node. Thus, if there are no compromised nodes and any two nodes can communicate with each other, any intermediate node will succeed in finding the next closer node v. Since each v has one more common sub-index than the corresponding u, the Hamming distance between each v and D will be smaller than that between u and D by 1. Therefore, there will be $d_h - 1$ intermediate nodes between S and D.

Lemma 3.3. *The number of intermediate nodes in the key path discovered in the above dynamic key path discovery algorithm never exceeds $2(d_h - 1)$.*

Proof. Consider the *flag* values in the sequence of unicast messages in the path discovery $\{flag_1, ..., flag_i\}$. First, it contains at most d_h ones since otherwise the request message should have already reached the destination node before the last message containing $flag = 1$. Second, there are no two consecutive zeros in this sequence since the second step will stop if it cannot find the next closer node and the flag in the current message is zero. Third, the last two values ($flag_{i-1}$ and $flag_i$) in this sequence are always 1 for a successful path discovery. Consider the last three nodes in the key path u, v, D, where D is the destination node. It is obvious that $flag_i$ is always 1 for a successful discovery. If $flag_{i-1} = 0$, the Hamming distance between u and D is 1, and there is no intermediate node between u and D. Thus, we know that both flag values are 1. Hence, we can conclude that the maximum length of this sequence is $d_h - 2 + d_h - 1 + 2 = 2d_h - 1$. This indicates that the maximum number of intermediate nodes in the key path is $2(d_h - 1)$.

3.4.3 Performance

Each sensor node stores n polynomial shares and each polynomial is shared by about m different nodes, where $m = \lceil \sqrt[n]{N} \rceil$. Thus, each node can establish direct keys with $n(m-1)$ other nodes. This indicates that the probability to establish direct keys between sensor nodes can be estimated by $\frac{n(m-1)}{m^n - 1} = \frac{n(\lceil \sqrt[n]{N} \rceil - 1)}{N-1}$. Figure 3.8(a) shows that the probability of establishing a direct key between two nodes decreases as the number of dimensions n or the network size N grows. However, according to the path discovery algorithm, if there are no compromised nodes and any two nodes can communicate with each other, it is guaranteed that any two nodes can establish a pairwise key (directly or indirectly).

3.4.4 Overhead

Each node has to store n polynomial shares and the IDs of revoked nodes with which it can establish direct keys. The former introduces $n(t + 1) \log q$ bits storage overhead. For the latter part, a node only needs to remember up to t compromised node IDs for each polynomial since if $t + 1$ shares of one bivariate polynomial are compromised, this polynomial is already compromised and can be discarded. In addition, a sensor node i only needs to remember one sub-index of each revoked ID because the IDs of node i and the revoked node only differ on one sub-index. Thus, at most ntl bits storage overhead is required to keep track of revoked node IDs. Totally, the storage overhead at senor nodes is at most $n(t+1) \log q + ntl$ bits, where $l = \lceil \log_2 m \rceil$. To establish a direct key, a sensor node only needs the ID of the other node, and there is no communication overhead. However, if two nodes cannot establish a direct key, they need to find a number of intermediate nodes to help them establish a temporary session key. When key path pre-determination is used, discovering a key path does not introduce any communication overhead. However, when dynamic key path discovery is used, this process involves a number of unicast messages. From Lemma 3.3, we know that the dynamic path discovery introduces at most $2d_h - 1$ unicast messages if every unicast message is successfully delivered.

Now let us estimate the communication overhead during the path key establishment, assuming the key path is already discovered. In sensor networks, sending a unicast message between two arbitrary nodes may involve the overhead of establishing a route. However, finding a route in a sensor network depends on routing protocol, which is also required by other schemes to do path discovery. In fact, we are unable to give a precise analysis on this overhead because of the undecided routing protocol. Thus, for simplicity, we use the number of unicast messages to estimate the communication overhead involved in the path key establishment. In fact, there are $L+1$ unicast messages for the path key establishment using a key path with length L if every unicast message is successfully delivered. If there are no compromise sensor nodes and any two nodes can communicate with each other, there exists at least one key path with $d_h - 1$ intermediate nodes, which is indeed one of the the the shortest key paths.

Consider two nodes u ($\langle u_1, ..., u_n \rangle$) and v ($\langle v_1, ..., v_n \rangle$). The probability of having $u_e = v_e$ for any $e \in \{1, \cdots, n\}$ is $\frac{1}{m}$, and the probability of having exactly i different sub-indexes is

$$P[i \text{ different sub-indexes}] = \frac{n!}{i!(n-i)!} \frac{1}{m^{n-i}} (1 - \frac{1}{m})^i.$$

Thus, the average key path length can be estimated by

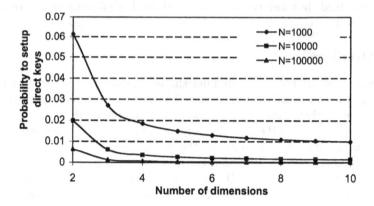

(a) Probability of establishing direct keys.

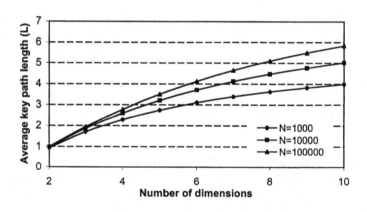

(b) Average key path length

Fig. 3.8. Performance of the hypercube-based scheme

$$L = \sum_{i=1}^{n}(i-1) \times P[i \text{ different sub-indexes}].$$

Figure 3.8(b) shows the relationship between the the average key path length and the number of dimensions given different network sizes. We can see that the average key path length (and thus the communication overhead) increases as the number of dimensions or the network size grows.

In terms of computational overhead, each unicast message requires one polynomial evaluation, one authentication and one encryption at the source node; and one polynomial evaluation, one authentication and one decryption at the destination node. Since $L + 1$ unicast messages are needed for a key path with length L, there are totally $2(L+1)$ polynomial evaluations, $2(L+1)$ authentications, $L+1$ encryptions, and $L+1$ decryptions involved in the path key establishment if every message is successfully delivered.

3.4.5 Security Analysis

An adversary may launch two types of attacks against the hypercube-based scheme. First, the attacker may target the pairwise key between two particular sensor nodes. The attacker may either try to compromise the pairwise key or prevent the two sensor nodes from establishing a pairwise key. Second, the attacker may target the entire network to lower the probability or to increase the cost to establish pairwise keys.

Attacks Against a Pair of Nodes

We focus on the attacks on the communication between the two particular sensor nodes u and v. Assume neither of them is compromised by the attacker. If these two nodes can establish a direct key, the only way to compromise the key without compromising the related nodes is to compromise the shared polynomial between them, which requires the attacker to compromise at least $t+1$ sensor nodes. If these two nodes have established an indirect key through path key establishment, the attacker may compromise one of the polynomials or the nodes involved in this establishment so that the attacker can recover the key if he has the message used to deliver this key. However, even if the attacker compromises the current pairwise key, the related sensor nodes may still re-establish another pairwise key with a different key path. To prevent u from establishing a pairwise key with v, the attacker has to compromise all those n polynomial shares on u (or v) so that node u (or v) is unable to use any polynomial to setup a pairwise key; otherwise, there may still exist non-compromised sensor nodes that can help establish a new pairwise key. For each of these polynomial shares, the attacker has to compromise at least $t+1$ nodes. This means that the attacker has to compromise at least $n(t+1)$ sensor nodes to prevent u and v from establishing another pairwise key.

Attacks Against the Network

Since the adversary also knows the distribution of polynomials over sensor nodes, it may systematically attack the network by compromising the polynomials in $\mathcal{F}$ one by one in order to compromise the entire network. Assume the attack compromises b bivariate polynomials. There are up to bm sensor nodes with at least one compromised polynomial share. Among all the remaining $N - bm$ sensor nodes, none of the secure links between them is compromised since the polynomials used to establish direct keys between them are not compromised. However, the indirect keys in the remaining part of the network could be affected since the common polynomial between two intermediate nodes in the key path might be compromised. Nevertheless, there is still a high probability of re-establishing a new indirect key between the two nodes even if an indirect key between two non-compromised nodes is compromised.

Alternatively, the adversary may randomly compromise sensor nodes to attack the path discovery process in order to make it more expensive to establish pairwise keys. In the following, we first investigate the probability of a direct key (secure link) being compromised, and then investigate the probability of any (direct or indirect) key being compromised under node compromises.

Assume a fraction p_c of sensor nodes in the network are compromised. Then, the probability that exactly i shares on a particular bivariate polynomial have been disclosed is

$$P[i \text{ compromised shares}] = \frac{m!}{i!(m-i)!}p_c^i(1-p_c)^{m-i},$$

where $m = \lceil \sqrt[n]{N} \rceil$. Thus, the probability of a particular bivariate polynomial being compromised is $P_{cd} = 1 - \sum_{i=0}^{t} P[i \text{ compromised shares}]$. If $m \gg t+1$, this is equivalent to the probability of any link (direct key) between non-compromised nodes being compromised. For a small m, P_{cd} only represents the fraction of compromised bivariate polynomials. For example, when $n = 4$ and $N = 20,000$, we have $m = 12$ and $t = 11$. In this case, we do not use the fraction of compromised bivariate polynomial to estimate the fraction of compromised links between non-compromised nodes. Instead, we note that the fraction of compromised links between non-compromised nodes in this situation is zero, which implies perfect security against node compromises.

Figure 3.9(a) shows the relationship between the fraction of compromised links for non-compromised nodes and the fraction of compromised sensor nodes with a different number of dimensions.[1] We can see that given the fixed network size and storage overhead, the hypercube-based scheme with more dimensions has higher security performance.

Now let us compute the probability of any (direct or indirect) key between two non-compromised nodes being compromised. Suppose sensor nodes u and v have different sub-indexes in i dimensions. The key path discovered between them involves $i-1$ intermediate nodes and i bivariate polynomials. If none of these $i-1$ intermediate nodes and i bivariate polynomials is compromised, the pairwise key is still secure; otherwise, this key cannot be trusted. This means that the probability of this pairwise key being compromised can be estimated by

---

[1] We assume there are totally $20,000$ sensor nodes in the following simulations. Thus, for a 4-dimensional hypercube, we have $m = \lceil \sqrt[4]{20,000} \rceil = 12$. This means that the degree of bivariate polynomial t is not necessarily larger than 11, and the sensor node needs to store at most $4 \times (11+1) = 48$ coefficients. Therefore, we assume that the storage constraint at sensor nodes is equivalent to storing 50 keys instead of 200 keys in the analysis of the earlier schemes.

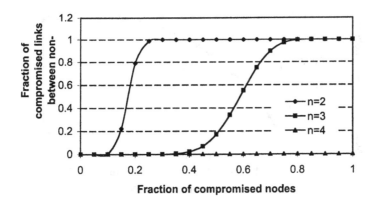

(a) Fraction of compromised links v.s. fraction of compromised nodes.

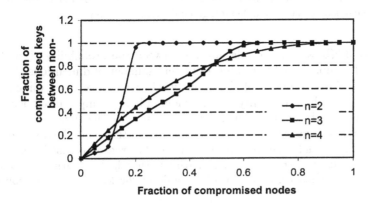

(b) Fraction of compromised direct and indirect keys v.s. fraction of compromised nodes.

Fig. 3.9. Security performance of the hypercube-based scheme. Assume each sensor
has available storage equivalent to 50 keys, $N = 20,000$, $m = \sqrt[n]{N}$, and $t = \lfloor \frac{50}{n} - 1 \rfloor$.

$$P[\text{comp.} \mid i \text{ different sub-indexes}] = 1 - (1 - p_c)^{i-1} \times (1 - P_{cd})^i.$$

Thus, the probability of any (direct or indirect) key between two non-
compromised nodes being compromised can be estimated by

$$P_c = \sum_{i=1}^{n} P[\text{comp.} \mid i \text{ different sub-indexes}] \times P[i \text{ different sub-indexes}].$$

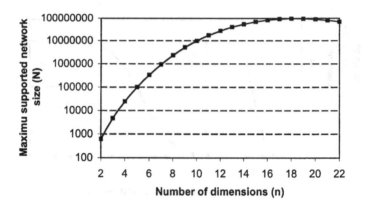

Fig. 3.10. Maximum supported network size for different number of dimensions. Assume each sensor has available storage equivalent to 50 keys.

Figure 3.9(b) shows the relationship between the probability P_c and the fraction of compromised nodes for a different number of dimensions. We can still see the improvements of the security when we have more dimensions. This is because the probability of a polynomial being compromised decreases quickly when we have more dimensions. We also note that when the fraction of compromised sensor nodes is less than a certain threshold, having more dimensions decreases the security of the scheme. The reason is that having more dimensions increases the average key path length, which in turn increases the probability of at least one intermediate node in the key path being compromised.

Maximum Supported Network Size

Let us consider the maximum supported network size when perfect security against node compromises is required. Figure 3.10 shows the maximum supported network size as a function of the number of dimensions given a fixed memory constraint and the guarantee of perfect security against node compromises. We can see that the maximum supported network size increases dramatically when we have more dimensions within the range shown in the figure. (Note that once the number of dimensions passes a certain threshold, this maximum supported network size will start to drop.) Indeed, when the number of dimensions is smaller, the hypercube-based scheme can support a larger network by adding more dimensions without increasing the storage overhead or sacrificing the security performance.

Probability of Re-Establishing a Pairwise Key

The following analysis estimates the probability of re-establishing an indirect key between two non-compromised nodes with the dynamic path discovery algorithm when all pre-determined key paths cannot be used due to compromised intermediate nodes or communication failures.

Let I_i denote the probability of establishing a pairwise key between two non-compromised nodes having different sub-indexes in i different dimensions (i.e., the Hamming distance between the two node IDs is $d_h = i$). For a particular node u, we refer to a non-compromised intermediate node as its *closer* node to the destination node if this node can establish a direct key with node u using a non-compromised polynomial and is closer to the destination node in terms of the Hamming distance between their IDs. According to the dynamic key path discovery algorithm, the pairwise key can be established if either of the following two cases is true. In the first case, the source node finds a closer node and the selected closer node finds a key path to the destination node. This probability can be estimated by

$$P_1 = \left[1 - [1 - (1 - P_{cd})(1 - p_c)]^i\right]I_{i-1}.$$

In the second case, the source node cannot find any closer node but can establish a direct key using a non-compromised polynomial with a non-compromised node that is able to find a closer node that can find a key path to the destination node. This probability can be estimated by

$$P_2 = (1 - P_1)(1 - P_{cd}^i)\left[1 - [1 - (1 - P_{cd})(1 - p_c)]^{i-1}\right]I_{i-1}.$$

Overall, we have $I_i = P_1 + P_2$ for $i > 1$ and $I_1 = 1 - P_{cd}$. Therefore, the probability of re-establishing an indirect key between two non-compromised nodes can be estimated by

$$P_{re} = \sum_{i=1}^{n} I_i \times P[i \text{ different sub-indexes}].$$

Figure 3.11 shows this probability for different fractions of compromised sensor nodes. It shows that even if a pairwise key is compromised, there is still a high probability of re-establishing a new key if the compromised nodes are detected. In addition, we note that the probability of re-establishing a key increases when there are more dimensions. This is because the probability of a polynomial being compromised decreases quickly as the number of dimensions grows.

3.4.6 Comparison with Previous Schemes

This subsection compares the hypercube-based key pre-distribution scheme with the basic probabilistic scheme [20], the q-composite scheme [12], the random pairwise keys scheme [12], and the random subset assignment scheme presented in Section 3.3. In the comparison, we use the hypercube-based scheme with $n = 2$ (i.e., the grid-based scheme) for simplicity.

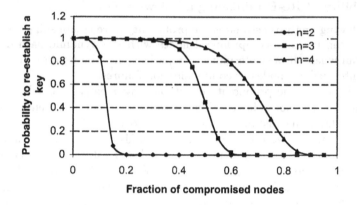

Fig. 3.11. Probability of re-establishing a pairwise key between non-compromised nodes v.s. the fraction of compromised nodes. Assume that each sensor node has available storage equivalent to 50 keys and $N = 20,000$.

The communication and computational overheads for direct key establishment in the grid-based scheme and the other schemes are summarized in Table 3.1. We can see that the grid-based scheme is generally much more efficient than the basic probabilistic scheme [20], the q-composite scheme [12], and the random subset assignment scheme in terms of the communication and computational overhead. Compared with the random pairwise keys scheme [12], the grid-based scheme involves only one more polynomial evaluation, which can be done very efficiently by using the optimization technique in Section 3.5.

To compare the security between different schemes, we assume that the network size is $N = 20,000$ and each node can store up to 200 keys or polynomial coefficients. In the grid-based scheme, we have $m = 142$ and $p = 0.014$. The four curves in the right part of figures 3.12(a) and 3.12(b) show the fraction of compromised links and the fraction of compromised (direct or indirect) keys between non-compromised nodes as a function of the number of compromised sensor nodes given $p = 0.014$. Similar to the comparison in Section 3.3, the random subset assignment scheme and the grid-based scheme perform much better when there are a small number of compromise nodes. In fact, these two scheme always have better performance when the number of compromised nodes is less than 14,000. When there are more than 14,000 compromised nodes, none of the schemes can provide sufficient security because of the large fraction of compromised links (over 60% compromised links) or the large fraction of compromised (direct or indirect) keys (over 90% compromised keys).

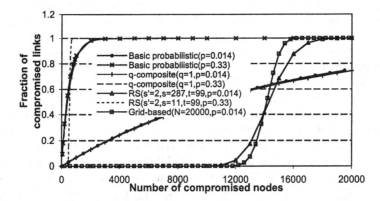

(a) Fraction of compromised links between non-compromised nodes v.s. number of compromised nodes.

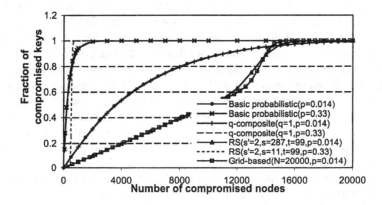

(b) Fraction of compromised (direct or indirect) keys between non-compromised nodes v.s. number of compromised nodes.

Fig. 3.12. Performance of the grid-based key pre-distribution scheme under attacks. Assume each sensor node has available storage equivalent to 200 keys.

Though $p = 0.014$ is acceptable for the grid-based scheme, for the basic probabilistic, the q-composite, and the random subset assignment schemes, p should be large enough to make sure the whole

network is fully connected. Assume that $p = 0.33$. This requires about 42 neighbor nodes for each sensor node to make sure the whole network with 20,000 nodes is connected with a high probability. The three curves in the left part of figures 3.12(a) and 3.12(b) show the fraction of compromised links and the fraction of compromised (direct or indirect) keys between non-compromised nodes as a function of the number of compromised sensors for the above three schemes when $p = 0.33$. We can see that a small number of compromised nodes reveal a large fraction of secrets in the network in these schemes; however, the fraction of compromised links and the fraction of compromised (direct or indirect) keys are much lower in the grid-based scheme for the same number of compromised nodes.

To compare with the random pairwise keys scheme [12], we set $m = t + 1$, so that the grid-based scheme can provide the same degree of perfect security guarantee as the random pairwise keys scheme. Assume that the storage overhead on sensor nodes is $2(t + 1) = 2m$. The grid-based scheme can support a network with m^2 nodes, and the probability that two sensor nodes share a direct key is $p = \frac{2}{m+1}$. With the same number sensor nodes and storage overhead, the random pairwise keys scheme [12] has $p = \frac{2m}{m^2} = \frac{2}{m}$, which is approximately the same as our scheme.

In addition to the above comparisons, the grid-based scheme has some unique properties that the other schemes do not provide. First, when there are no compromised sensor nodes in the network, it is guaranteed that any pair of sensor nodes can establish a pairwise key either directly without communication or through the help of an intermediate node when the sensor nodes can communicate with each other. Besides the efficiency in determining the key path, the communication overhead is substantially lower than the previous schemes which require real-time path discovery even in normal situations. Second, even if there are compromised sensor nodes in the network, there is still a high probability that two non-compromised sensor nodes can re-establish a pairwise key. Our earlier analysis indicates that it is very difficult for the adversary to prevent two non-compromised nodes from establishing a pairwise key. Finally, due to the orderly node assignment, this scheme allows optimized deployment of sensor nodes so that the sensor nodes that can establish direct keys are close to each other, thus greatly decreasing the communication overhead in path key establishment.

3.5 Implementation and Evaluation

We have implemented the random subset assignment scheme and the grid-based scheme [2] on MICA2 motes [14] running TinyOS [26], which is an operating system for networked sensor nodes. These implementations were written in nesC [21], a C-like programming language used to develop TinyOS and its applications. A critical component in our implementations is the algorithm to evaluate a t-degree polynomial, which is used to compute pairwise keys. In the following, we first present an optimization technique for polynomial eval-

---

[2] These implementations are included in our *tiny key management* (TinyKey-Man) package, which is available online at http://discovery.csc.ncsu.edu/software/TinyKeyMan.

uation on sensor nodes and then report the evaluation of this optimization technique and our key pre-distribution schemes.

3.5.1 Optimization of Polynomial Evaluation

Evaluating a t-degree polynomial is essential in the computation of a pairwise key in our schemes. This requires t modular multiplications and t modular additions in a finite filed F_q, where q is a prime number that is large enough to accommodate a cryptographic key. This implies that q should be at least 64 bits long for typical cryptosystems such as RC5. However, processors in sensor nodes usually have a much smaller word size. For example, ATmega128, which is used in many types of sensors, only supports 8-bit multiplications and has no division instruction. Thus, in order to use the basic scheme, sensor nodes have to implement some large integer operations.

Nevertheless, in our schemes, polynomials can be evaluated in much cheaper ways than polynomial evaluation in general. This is mainly due to the observation that the points at which the polynomials are evaluated are sensor IDs, and these IDs can be chosen from a different finite field $F_{q'}$, where q' is a prime number that is larger than the maximum number of sensors but much smaller than a typical q.

During the evaluation of a polynomial $f(x) = a_t x^t + a_{t-1} x^{t-1} + \cdots + a_0$, since the variable x is the ID of a sensor, the modular multiplication is always performed between an integer in F_q and another integer in $F_{q'}$. For example, to compute the product of two 64-bit integers on an 8-bit CPU, it takes 64 word multiplications with the standard large integer multiplication algorithm and 27 word multiplications with the Karatsuba-Ofman algorithm [36]. In contrast, it only takes 16 word multiplications with the standard algorithm to compute the product of a 64-bit integer and a 16-bit integer on the same platform. Similarly, reduction of the later product (which is an 80-bit integer) modulo a 64-bit prime is also about 75% cheaper than the former product (which is a 128-bit integer).

Considering the lack of a division instruction in typical sensor processors, we further propose to use q' in the form of $q' = 2^k + 1$. Because of the special form of $q' = 2^{16} + 1$, no division operation is needed to compute modular multiplications in $F_{q'}$ [76]. Two natural choices of such prime numbers are $257 = 2^8 + 1$ and $65,537 = 2^{16} + 1$. Using the random subset assignment scheme, these two special finite fields can accommodate up to 256 and 65,536 sensors, respectively; using the grid-based scheme, these two special finite fields can accommodate up to $256^2 = 65,536$ and $65,536^2 = 4,294,967,296$ sensors, respectively.

To take full advantage of the special form of q', we propose to adapt the basic polynomial-based key pre-distribution in Section 3.1.1 so that a large key is split into pieces and each piece is distributed to sensors with a polynomial over $F_{q'}$. The same technique can be easily applied to all polynomial pool-based schemes with slight modification.

Assume that each cryptographic key is n bits. The setup server divides the n-bit key into r pieces of l-bit segments, where $l = \lfloor \log_2 q' \rfloor$ and $r = \lceil \frac{n}{l} \rceil$. For simplicity, we assume that $n = l \cdot r$. The setup server randomly generates r t-degree bivariate polynomials $\{f_v(x, y)\}_{v=1,\cdots,r}$ over $F_{q'}$ such that $f_v(x, y) = f_v(y, x)$ for $v = 1, \cdots, r$. The setup server then gives the corresponding polynomial shares on these r polynomials to each sensor node. Specifically, each sensor node i receives $\{f_v(i, x)\}_{v=1,\cdots,r}$. With the basic scheme, each of these r polynomials can be used to establish a common secret between a pair of sensors. These sensors then choose the l least significant bits of each secret value as a key segment. The final pairwise key can simply be the concatenation of these r key segments.

It is easy to verify that this method requires the same number of word multiplications as the earlier one; however, because of the special form of q', no division operation is necessary in evaluating the polynomials. This can significantly reduce the computation on processors that do not have any division instruction.

The security of this scheme is guaranteed by Lemma 3.4.

Lemma 3.4. *In the adapted key-pre-distribution scheme, the entropy of the key for a coalition of no more than t other sensor nodes is $r \cdot [\log_2 q' - (2 - \frac{2^{l+1}}{q'})]$, where $l = \lfloor \log_2 q' \rfloor$ and $r = \lceil \frac{n}{l} \rceil$.*

Proof. Assume that nodes u and v need to establish a pairwise key. Consider a coalition of no more than t other sensor nodes that tries to determine this pairwise key. According to the security proof of the basic key pre-distribution scheme [6], the entropy of the shared secret derived with any polynomial is $\log q'$ for the coalition. That is, any value from the finite field $F_{q'}$ is a possible value of each of $\{f_j(u, v)\}_{j=1,\dots,r}$ for the coalition. Since each piece of key consists of the last $l = \lfloor \log_2 q' \rfloor$ bits of one of the above values, values from 0 to $q' - 2^l - 1$ have the probability $\frac{2}{q'}$ to be chosen, while the values from $q' - 2^l$ to $2^l - 1$ have the probability $\frac{1}{q'}$ to be chosen. Denote all the information that the coalition knows as C. Thus, for the coalition, the entropy of each piece of key segment K_j, $j = 1, \cdots, r$, is

$$H(K_j|C) = \sum_{i=0}^{q'-2^l-1} \frac{2}{q'} \log_2 \frac{q'}{2} + \sum_{i=q'-2^l}^{2^l-1} \frac{1}{q'} \log_2 q'$$

$$= \frac{2(q' - 2^l)}{q'} \log_2 \frac{q'}{2} + \frac{2^{l+1} - q'}{q'} \log_2 q'$$

$$= \log_2 q' - (2 - \frac{2^{l+1}}{q'})$$

Because the r pieces of key segments are distributed individually and independently, the entropy of the pairwise key for the coalition is

$$H(K|C) = \sum_{j=1}^{r} H(K_j|\cdot) = r \cdot [\log_2 q' - (2 - \frac{2^{l+1}}{q'})].$$

Consider a 64-bit key. If we choose $q' = 2^{16} + 1$, the entropy of a pairwise key for a coalition of no more than t compromised sensor nodes is $4 \times [\log_2(2^{16} + 1) - (2 - \frac{2^{17}}{2^{16}+1})] = 63.9997$ bits. If we choose $q' = 2^8 + 1$, this entropy is then $8 \times [\log_2(2^8 + 1) - (2 - \frac{2^9}{2^8+1})] = 63.983$ bits. Thus, the adapted scheme still provides sufficient security despite the minor leak of information.

3.5.2 Evaluation

We first evaluate the performance of our optimization technique for polynomial evaluation. This optimization forms the basis of pairwise key computation in our implementation. We provide two options for this component: one with $q' = 2^8 + 1$ and the other with $q' = 2^{16} + 1$. The typical length of a cryptographic key in resource constrained sensor nodes is 64 bits. To compute a 64-bit pairwise key, a sensor node has to evaluate 8 t-degree polynomials if $q' = 2^8 + 1$ and 4 t-degree polynomials if $q' = 2^{16} + 1$. The code sizes for the implementations of these two options are shown in Table 3.2. The bytes needed for polynomial coefficients are not included in the code size calculation since it depends on the applications. Obviously, these two implementations occupy only a small amount of memory at sensor nodes.

Table 3.2. Code sizes for our optimized polynomial evaluation schemes.

Scheme	ROM (bytes)	RAM (bytes)
$q' = 2^8 + 1$	288	11
$q' = 2^{16} + 1$	416	20

The cost of our optimization technique in computing a 64-bit cryptographic key on a MICA2 mote [14] is shown in Figure 3.13, which also includes the cost of generating a 64-bit MAC (Message Authentication Code) for a 64-bit long message using RC5 [67] and SkipJack [58] with a 64-bit long key. These two symmetric cryptographic techniques are generally believed to be practical and efficient for sensor networks. The result shows that computing a pairwise key in our schemes can be faster than generating a MAC using RC5 or SkipJack for a reasonable polynomial degree t; and in practice, it is not necessary for the value of t to be a very large number due to the storage and security concerns, which can be seen from previous analysis in sections 3.3 and 3.4. The result demonstrates the practicality and efficiency of our proposed schemes.

According to the result in Figure 3.13, the 16-bit version is slightly slower than the 8-bit version. However, the 16-bit version can accommodate a lot

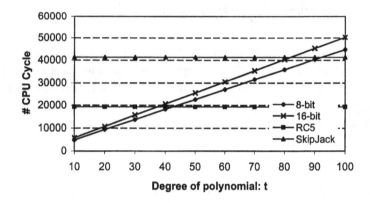

Fig. 3.13. Comparison with RC5 and SkipJack.

Table 3.3. The code size for random subset assignment and grid-based scheme. The storage for the polynomial coefficients and the list of compromised nodes are not included in the calculation of code size.

Scheme	ROM (bytes)	RAM (bytes)
Random Subset Assignment	2824	106
Grid-Based	1160	67

more sensor nodes than the 8-bit version. Thus, we use the 16-bit option for both the random subset assignment scheme and the grid-based scheme. The code sizes for these two schemes are shown in Table 3.3, which only includes the size of code loaded on sensor nodes, since the components for the setup server are not installed on sensor nodes. In fact, the setup server is not necessary to be a sensor node. We can see that the code size for the grid-based scheme is much smaller than that for the random subset assignment scheme since the grid-based scheme can directly determine the direct key shared or the key path involved; while the random subset assignment scheme has to contact other nodes and maintain many more states.

Considering the analysis in previous sections and the evaluation of computational and storage cost, we can conclude that our schemes are practical and efficient for the current generation of sensor networks.

3.6 Summary

In this chapter, we developed a general framework for polynomial pool-based pairwise key pre-distribution in sensor networks based on the basic polynomial-based key pre-distribution in [6]. This framework allows study of multiple instantiations of possible pairwise key establishment schemes. Two possible instantiations developed were the key pre-distribution scheme based on random subset assignment and the hypercube-based key pre-distribution scheme. Our analysis demonstrated that our schemes have significant advantages over the existing approaches. The experiment results also demonstrate its practicality and efficiency in real sensor networks.

As one of possible future directions, we observe that sensor nodes have low mobility in many applications. Thus, it may be desirable to develop location-based schemes so that the nodes that can directly establish a pairwise key are close to each other. In addition, the grid-based scheme or the hypercube-based scheme can also be extended to a more general scheme which has a different number of bivariate polynomials arranged in different dimensions.

4

Improving Pairwise Key Establishment Using Deployment Knowledge

In this chapter, we first exploit the *prior deployment knowledge* of sensor nodes to improve key pre-distribution in static sensor networks. The techniques are based on the observation that in static sensor networks, *although it is difficult to precisely pinpoint sensor nodes' positions, it is often possible to approximately determine their locations*. For example, when trucks are used to deploy static sensor nodes, the sensor nodes can usually be kept within a certain distance (e.g., 100 yards) from their target locations – even though it is difficult to place the sensor nodes in their expected locations precisely. By taking advantage of this observation, our techniques provide better security and performance than the previous techniques.

We then propose to take advantage of *post deployment knowledge* and investigate a new approach, which we refer to as *key prioritization* , to improve the performance of key pre-distribution schemes in static sensor networks. The main idea is to use the memory for applications (e.g., EEPROM on MICA2 motes [14]) to store an excessive amount of keying materials, prioritize the keying materials based on sensor nodes' post deployment information, and discard low priority keying materials to thwart node compromise attacks, as well as return memory to the applications. For example, a sensor application may be designated to collect temperature, humidity, etc. at a certain frequency and buffer the data before transmitting it back to the central processing system. During the key pre-distribution phase (before the deployment of the sensor nodes), we may use a large amount of available memory (e.g., in EEPROM) to store pre-distributed keying materials. After a sensor node is deployed, it may first examine its environment to assess the likelihood of using each keying material, prioritize the keying materials accordingly, and then discard the low priority ones.

We also develop a group-based key pre-distribution framework when the locations of sensor nodes cannot be pre-determined or discovered after deployment. Compared to the previous techniques for improving key pre-distribution, this approach has the following two advantages. First, it does not require any prior knowledge of sensors' locations, which greatly simplifies the deployment of sensor networks. Second, the proposed framework can be easily combined with any of the existing key pre-distribution techniques, while previous techniques can only be used to improve a certain type of key pre-distribution techniques. The analysis indicates that the framework improves the security as well as the performance of existing key pre-distribution techniques substantially.

4.1 Improving Key Pre-Distribution with Prior Deployment Knowledge

In static sensor networks, it may be possible to predetermine the locations of sensor nodes to a certain extent. These predetermined locations can be used to improve the performance of pairwise key pre-distribution. In this section, we first introduce a simple location-aware deployment model for this purpose, and then develop two pairwise key pre-distribution schemes that can take advantage of the predetermined location information.

4.1.1 A Location-Aware Deployment Model

We assume that sensor nodes are deployed in a two dimensional area called *target field* , and two sensor nodes can communicate with each other if they are within each other's *signal range* . The location of a sensor node can be represented by a coordinate in the target field. Each sensor node has an *expected location* that can be predicted or predetermined. After the deployment, a sensor node is placed at a *deployment location* that may be different from its expected location. We call the difference between the expected location and the deployment location of a sensor node the *deployment error* for this sensor node. This deployment model can be characterized by the following three parameters:

1. **Signal range** d_r**:** A sensor node can receive messages from another sensor node if the former is located within the signal range of the latter. We model the signal range of a sensor node as a circle centered at its deployment location with the radius d_r. For simplicity, we assume that the radius d_r defining the signal range is a network-wide parameter, and denote the signal range with d_r. We say two sensor nodes are *neighbors* if they are physically located within each other's signal range.

2. **Expected location** (L_x, L_y)**:** The expected location (L_x, L_y) of a sensor node is a coordinate in the two dimensional target field; it specifies where the sensor node is expected to be deployed. Sometimes, a sensor node may be expected to be deployed within an area instead of a particular location. In this case, we assume that the sensor node is expected to be deployed at any location in that area with equal probability.

3. **Deployment pdf ϵ:** We model the actual deployment location of a sensor node with a *probability density function* ϵ. The sensor node expected to be deployed at (L_x, L_y) may appear at a particular area with a certain probability, which is calculated by the integration of a probability density function ϵ over this area. In some cases, the sensor node may have certain mobility and appear somewhere near its expected location with a certain probability. The deployment location of this sensor node at any point in time may also be modeled by the probability density function.

Although our techniques can be applied to any deployment model, we always evaluate the performance of our techniques with a simple model, where each sensor node randomly appears anywhere at a distance of no more than e away from the expected location. We call e the *maximum deployment error*. Thus, the deployment pdf ϵ for a sensor node u with expected location (L_x, L_y) can be expressed as

$$\epsilon_{(L_x, L_y)}(x, y) = \begin{cases} \frac{1}{\pi e^2}, & ||(L_x, L_y), (x, y)|| \le e \\ 0, & otherwise. \end{cases}$$

where $|| \cdot ||$ denotes the distance between two locations.

Obviously, this model can be easily extended to a three dimensional space. However, we focus on pairwise key establishments in the two dimensional case here. Extending our results to the three dimensional model would be straightforward.

4.1.2 Closest Pairwise Keys Scheme

In this subsection, we develop a pairwise key establishment scheme named *closest pairwise keys scheme* to take advantage of the expected location information. The basic idea is to have each sensor node share pairwise keys with a number of other sensor nodes whose expected locations are closest to the expected location of this sensor node. The following discussion starts with a basic version, which can be considered as the combination of the random pairwise keys scheme [12] and the expected location information, and then gives an extended version to further reduce the storage overhead and facilitate dynamic deployment of new sensor nodes.

We assume a setup server is responsible for key pre-distribution. This setup server is aware of the expected location of each sensor node. However, it does not require the network-wide signal range d_r and the deployment pdf ϵ since these two pieces of information are not used in our technique. We assume each sensor node has a unique, integer-valued ID. We also use this node ID to refer to the sensor node. For convenience, we call a pairwise key shared directly between two neighbor nodes a *direct key* and a pairwise key established through other intermediate nodes an *indirect key*.

The Basic Version

The basic idea of the closest pairwise keys scheme is to *pre-distribute pairwise keys between pairs of sensor nodes that have high probabilities to be neighbors.* Though reasonable, this idea is difficult to implement since it is non-trivial to determine the probability that two sensor nodes are neighbors. Indeed, this probability depends on the deployment pdf ϵ, which is generally not available and may vary in different applications. To simplify the situation, we pre-distribute pairwise keys between pairs of sensor nodes whose expected locations are close to each other, hoping that the closer the expected locations of two sensor nodes, the more likely they are to be physically located in each other's signal range.

1. **Pre-Distribution.** Based on the expected locations of sensor nodes, the setup server pre-distributes pairwise keys for each sensor node to facilitate the pairwise key establishment during the normal operation. Specifically, for each sensor node u, the setup server first discovers a set S of c other sensor nodes whose expected locations are closest to the expected location of u – where c is a system parameter determined by the memory constraint. For each node v in S, the setup server randomly generates a unique pairwise key $K_{u,v}$ if no pairwise key between u and v has been assigned. The setup server then assigns $(v, K_{u,v})$ and $(u, K_{u,v})$ to sensor nodes u and v, respectively.

2. **Direct Key Establishment.** After the deployment of the sensor network, if two sensor nodes u and v want to establish a pairwise key to secure the communication between them, they only need to check whether there is a pre-distributed pairwise key between them. This information is obtained from the setup server in the pre-distribution phase. The algorithm to identify such a common key is trivial because each pairwise key in a particular sensor node is associated with a node ID.

3. **Indirect Key Establishment.** When two neighbor nodes cannot establish a direct key, they need to find one or more intermediate nodes to help them setup an indirect session key. A simple way is to have one node (*called source node*) send a request to a number of nodes that share direct keys with it. If one of those contacted nodes also shares a direct key with the other node (*called destination node*) , this contacted node can be used as an intermediate node to help establish a common session key. In our later schemes, we will omit the indirect key establishment phase since it is not directly related to our techniques. Indeed, indirect key establishment can be done with any of the previous schemes (e.g., [42, 12, 20]).

4. **Sensor Addition and Revocation.** During the lifetime of a sensor network, new sensor nodes may be added to replace damaged or compromised sensor nodes. To add a new node, the setup server performs the above pre-distribution process for the new sensor node and then informs the deployed sensor nodes chosen for the new sensor node the corresponding pairwise keys through secure channels. (Here we assume the commu-

nication between each sensor node and the setup server is secured with a unique pairwise key shared between the node and the setup server.) The setup server may know the deployment locations of the deployed sensor nodes. In this case, the setup server may use these deployment locations (instead of their expected locations) to select neighbors for the new sensor node.

The detection of compromised nodes is generally a difficult problem, which is beyond our scope. However, there are several methods that could be used to identify compromised sensor nodes to revoke (e.g., [49, 8, 19, 47]). Once the compromised nodes are detected, it is usually necessary to revoke them from the network. To revoke a sensor node, each sensor node that shares a pairwise key with the revoked node simply deletes the corresponding key from its memory.

Though the above scheme looks similar to the previous methods [20, 12], it can achieve better performance if the predetermined location information is available. In the following, we show the improvement over the previous methods through analysis. Table 4.1 lists several notations that are often used in our analysis.

Table 4.1. Notations

N	network size
m	average number of neighbor sensor nodes
c	number of keys pre-distributed to sensor nodes
p	probability of sharing a direct key between neighbors
P_c	fraction of compromised direct keys between non-compromised nodes

Probability of Establishing Direct Keys

For simplicity, we assume that the sensor nodes in the network are expected to be evenly distributed in the target field. Thus, if u is one of v's closest c sensor nodes, v is very likely to be one of u's closest c sensor nodes. We use $\dot{u}$ and $\bar{u}$ to represent the deployment location and the expected location of node u, respectively. As discussed in Section 4.1.1, we model the deployment location of node u as a probability density function $\epsilon_{\bar{u}}(x, y)$.

Consider two sensor nodes u and v. Since they are deployed independently, given the expected locations of u and v, the conditional probability that they are neighbors can be calculated by

$$p(||\dot{u}, \dot{v}|| \leq d_r | \bar{u}, \bar{v}) = \int \int \int \int_{||\dot{u}, \dot{v}|| \leq d_r} \epsilon_{\bar{u}}(x_1, y_1) \epsilon_{\bar{v}}(x_2, y_2) dx_1 dy_1 dx_2 dy_2,$$

where $|| \cdot ||$ denotes the distance between two locations.

Since sensor nodes are evenly distributed in the target field, the densities of sensor nodes in different small areas are approximately equal. Assume there are on average m nodes in each sensor node's signal range. The density of the network can be estimated by $D = \frac{m+1}{\pi d_r^2}$, where d_r is the radius of the signal range. Thus, on average, each node will get pre-distributed pairwise keys with the sensor nodes whose expected locations are no more than d' away from it, where $d' = \sqrt{\gamma} \times d_r$, and $\gamma = \frac{c}{m+1}$. We call $\gamma = \frac{c}{m+1}$ the *capacity density ratio*. For any v having a pre-distributed pairwise key with u, the probability that v falls into u's signal range can be calculated by

$$p(||\dot{u}, \dot{v}|| \leq d_r | \bar{u}) = \int \int_{||\bar{u}, \bar{v}|| \leq d'^2} \frac{p(||\dot{u}, \dot{v}|| \leq d_r | \bar{u}, (x, y))}{\pi d'^2} dx dy.$$

Among the sensor nodes that have pre-distributed pairwise keys with sensor node u, the average number of sensor nodes that fall into its signal range can be estimated by $c \times p(||\dot{u}, \dot{v}|| \leq d_r | \bar{u})$. Thus, the probability of establishing a common key between node u and its neighbor sensor node can be estimated by

$$p = \frac{c \times p(||\dot{u}, \dot{v}|| \leq d_r | \bar{u})}{m} \approx \gamma \times p(||\dot{u}, \dot{v}|| \leq d_r | \bar{u}).$$

The above p can usually be used to estimate the probability of any node having a direct key with its neighbor node when the target field is infinite. For a limited field in our simulation, we simply use the probability p of the node expected to be deployed at the center of this field having a direct key with its neighbor node to estimate the probability of having a direct key between any two neighbor nodes.

In the following analysis, we always use the radius of signal range, d_r, as the basic unit of distance measurement ($d_r = 1$). For example, a distance 2 implies that the distance is twice as far as d_r. Thus, the deployment error in our discussion also represents the ratio of the deployment error to the signal range. Figure 4.1 shows the probability of establishing direct keys between neighbor sensor nodes for different values of e and γ. We can see that this probability is not only affected by the maximum deployment error, but also by the capacity density ratio γ. In general, the increase of γ will increase the probability p given certain deployment pdf. However, this probability decreases when the maximum deployment error increases. In practice, we expect to see better performance than that in Figure 4.1 since the probability of having a smaller deployment error is typically higher than the probability of having a larger one.

Security Against Node Captures

There are several attacks against sensor networks, such as DoS attacks [79], Sybil attacks [53], and Wormhole attacks [29]. These attacks may affect the

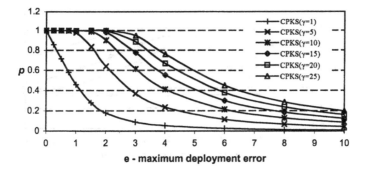

Fig. 4.1. Probability of establishing direct keys between two neighbor nodes given different values of e and γ. CPKS denotes the closest pairwise keys pre-distribution scheme.

security of pairwise key establishment in sensor networks. For example, an attacker may create wormholes between different areas so that a node establishes unnecessary keys with other nodes that will disappear once the wormholes are gone. However, these attacks are not unique to the pairwise key establishment in sensor networks. In addition, using expected location information does not reduce the security of any existing key pre-distribution scheme. For simplicity, we focus on the node compromise attacks in the following.

In node compromise attacks, an adversary may physically capture one or more sensor nodes and learn all the secrets stored on these nodes. We assume the compromised sensor nodes may collude together to attack the communication between non-compromised sensor nodes. That is, the adversary may try to figure out the pairwise keys established between non-compromised nodes based on the secrets learned from the compromised ones.

From the scheme it is easy to see that each pre-distributed pairwise key between two sensor nodes is randomly generated. Thus, no matter how many sensor nodes are compromised, the direct keys between non-compromised sensor nodes are still secure. We call this property *perfect security against node captures* . However, once a sensor node is compromised, the session keys that this sensor node helps establish may be compromised. For example, an attacker may have saved a copy of an indirect session key encrypted by a direct key and thus will be able to decrypt the session key once she gets the corresponding direct key from a compromised node. Thus, if either of the source or destination sensor nodes notices that the intermediate sensor node is compromised, it should remove the corresponding pre-distributed pairwise key and initiate a request to establish a new session key. The delay in detecting compromised sensor nodes still poses a threat. One way to mitigate this threat is

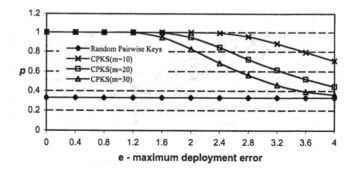

Fig. 4.2. Probability of establishing direct keys in random pairwise keys scheme and the closest pairwise keys scheme for different m and e given $c = 200$ and $N = 600$.

to derive the session key by combining (e.g., XOR) the keys generated from multiple paths, as suggested in [12].

Overhead

Ideally, each sensor node stores c pairwise keys. However, this does not necessarily happen because of the asymmetry in sensor nodes' locations. Consider a pair of sensor nodes u and v. In the pre-distribution step, v is one of u's closest c sensor nodes; however, u is not necessarily one of v's closest c sensor nodes. In this case, v has to store the pairwise key between u and v in addition to its own pre-distributed c pairwise keys. Thus, the storage overhead in each sensor node comes from two parts: one consists of the pairwise keys generated for itself, and the other consists of the pairwise keys generated for other sensor nodes. Hence, each sensor node has to store at least c keys and c sensor IDs. The actual number of pairwise keys stored in a particular sensor node may be much larger than c. Nevertheless, if the sensor nodes are approximately evenly distributed in the target field, it is very likely that if sensor node u is among sensor node v's closest c sensor nodes, then v is among u's closest c sensor nodes.

To establish a common key with a given neighbor node, a sensor node only needs to check whether it has a pre-distributed pairwise key with the given node (because each pairwise key is associated with a node ID). Thus, there is no communication and computation overhead during direct key establishment. The establishment of an indirect session key requires one broadcast request message, and potentially a number of unicast reply messages with the techniques in [20, 12, 42].

Improvements

Our basic scheme can be considered as an extension to the random pair-wise keys scheme. These two schemes have some common properties. In both schemes, the compromise of sensor nodes does not lead to the compromise of direct keys shared between non-compromised sensor nodes. However, our scheme further takes advantage of expected location information and thus is able to achieve better performance than the random pairwise keys scheme. First, the random pairwise keys scheme has a restriction on the network size, while our scheme has no direct restriction on the network size. Second, given the same storage capacity c for pairwise keys and the total number N of sensor nodes, the probability of establishing direct keys in our scheme is always better than the random pairwise keys scheme. This is illustrated in Figure 4.2, which compares the probability of establishing direct keys in both schemes for different m and e given that $c = 200$ and $N = 600$. It shows that the probability p of establishing direct keys is improved significantly in our scheme, especially when e is less than two times of the signal range. When the maximum deployment error e increases, this probability gradually decreases. In the extreme case, when there is no knowledge about where the nodes may reside, the technique degenerates into the original random pairwise keys scheme.

Comparison with Previous Methods

Now let us compare our scheme with the basic probabilistic scheme [20], the q-composite scheme [12], and the random subset assignment scheme [42]. As discussed earlier, our proposed scheme has a high probability to establish direct keys between neighbor sensor nodes given reasonable capacity density ratio γ and maximum deployment errors. At the same time, our scheme does not put any limitation on the network size.

Note that the closest pairwise keys scheme provides perfect security against node capture attacks, while the basic probabilistic scheme and the q-composite scheme cannot achieve the perfect security guarantee. Although the random subset assignment scheme can be configured to achieve the perfect security guarantee, it can only support a limited number of sensor nodes to ensure a certain probability of having direct keys between sensor nodes. Thus, a more direct and reasonable way to make comparisons between different schemes is to show the security against node compromise attacks given the same probability of establishing direct keys between sensor nodes.

Figure 4.3 shows that given the same storage overhead and the same probability of establishing direct keys between sensor nodes, our scheme does not lead to compromise of direct keys belonging to non-compromised sensor nodes, while in the other three schemes [20, 12, 42], the direct keys shared between non-compromised sensor nodes are compromised quickly when the number of compromised sensor nodes increases.

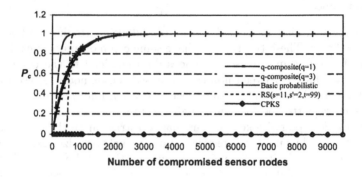

Fig. 4.3. Fraction of compromised pairwise keys between non-compromised sensor nodes v.s. number of compromised sensor nodes. RS denotes the random subset assignment scheme

Compared with the grid-based scheme [42], the closest pairwise keys scheme has some advantages. First, the closest pairwise keys scheme provides the perfect security guarantee. Though the grid-based scheme can also provide perfect security guarantee, it has limitations on the maximum supported network size given a certain storage constraint [42]. In contrast, the closest pairwise keys scheme has no limitations on the maximum supported network size. Second, the closest pairwise keys scheme can achieve a higher probability of establishing direct keys between two neighbor nodes than the grid-based scheme. Though the grid-based scheme can guarantee to establish a direct or indirect key between any two sensor nodes, it requires that any two sensor nodes can communicate with each other, which may not be true in real applications.

The Extended Version

The basic scheme described above has two limitations. First, if the sensor nodes are not evenly distributed in the target field, it is possible for a sensor node to have a large number of neighbor nodes that are not among the closest c sensor nodes of u but consider u as one of their closest c sensor nodes. As a result, node u has to store a lot of pairwise keys generated by the setup server. Second, to add a new sensor node after deploying the sensor network, the setup server has to inform a number of existing sensor nodes in the network about the addition of the new sensor node, which may introduce a lot of communication overhead.

We propose an alternative way to pre-distribute the secret information so that (1) the storage overhead in each node is small and fixed no matter how the sensor nodes are deployed, and (2) no communication overhead is introduced during the addition of new sensor nodes. The technique is based on a pseudo random function (PRF) [23] and a master key shared between each sensor node and the setup server.

1. **Pre-Distribution:** For each sensor u, the setup server first randomly generates a master key K_u and determines a set S of c other sensor nodes whose expected locations are closest to that of u. The setup server then distributes to sensor u a set of pairwise keys (together with the IDs) with those selected sensor nodes. These keys are generated by the setup server in the following way: For each $v \in S$, the setup server generates a pseudo random number $k_{u,v} = PRF_{K_v}(u)$ as the pairwise key shared between u and v, where K_v is the master key for v. As a result, for each $v \in S$, node u stores the pairwise key $k_{u,v}$, while node v can compute the same key with its master key and the ID of node u. We call v a *master sensor node* of u if the direct key $k_{u,v}$ shared between them is derived by $k_{u,v} = PRF_{K_v}(u)$. Accordingly, we call u a *slave sensor node* of v if v is a master sensor node of u.

2. **Direct Key Establishment:** The direct key establishment stage is similar to the basic scheme. The only difference is that one of two sensor nodes has a pre-distributed pairwise key and the other only needs to compute the key using its master key and the ID of the other party. For example, if u finds that it has the pre-distributed pairwise key $PRF_{K_v}(u)$ with v, it then notifies sensor v that it has such a key. Sensor node v only needs to compute $PRF_{K_v}(u)$ by performing a single pseudo random function.

3. **Sensor Addition and Revocation:** To add a new sensor node u, the setup server selects c sensor nodes closest to the expected location of u. For each of these c sensor nodes, the setup server retrieves v's master key K_v and computes $k_{u,v} = PRF_{K_v}(u)$, and then distributes v and $k_{u,v}$ to u. Revoking a sensor is a little more complex than in the basic scheme. To revoke sensor v, all its slave sensor nodes need to remove the corresponding keys from their memory. Moreover, v's master sensor nodes have to remember v's ID in order to avoid establishing a direct key with v later.

In this extension, each sensor node needs to store a master key which is shared with the setup server and c pre-distributed pairwise keys. Thus, the storage overhead for keys in each sensor node is at most $c + 1$. To establish a pairwise key, one of them can initiate a request by informing the other party that it has the pre-distributed pairwise key. (Note that this message only indicates the existence of such a key, and the exact value is never disclosed in the communication channel.) Once the other party receives such a message, it can immediately compute the pairwise key by performing one PRF operation. Thus, the communication overhead in the above scheme involves only one short request message, and the computation overhead only involves one efficient PRF operation.

Based on the security of PRF [23], if a node's master key is not disclosed, no matter how many pairwise keys generated with this master key are disclosed, it is still computationally infeasible for an attacker to recover the master key and the non-disclosed pairwise keys generated with different IDs. Thus, node compromise does not lead to the compromise of the direct keys shared between non-compromised nodes.

The extended scheme introduces some additional overhead by requiring master sensor nodes to remember the IDs of their revoked slaves. We consider this an acceptable overhead due to the following reasons. First, the storage overhead for a node ID is much smaller than that for one cryptographic key. Second, in normal situations when authentication of the revocation information is ensured, the number of revoked slave nodes is usually less than m, the average number of sensor nodes in each sensor's signal range. One may argue that if the authentication of revocation information can be bypassed, an attacker may convince a sensor node to store many node IDs to exhaust its memory. However, in this case, the node can be convinced to do anything and should be considered compromised.

4.1.3 Closest Polynomials Pre-Distribution Scheme

The scheme presented earlier still has some limitations. In particular, given the constraints on the storage capacity, node density, signal range and deployment pdf, the probability of establishing direct keys is fixed. For a particular sensor network, it is not convenient to adjust the last three parameters. Thus, one has to increase the storage capacity for pairwise keys to increase the probability of establishing direct keys. This may not be a feasible solution in certain sensor networks given the memory constraints on sensor nodes.

This subsection presents a key pre-distribution scheme, called *closest polynomials pre-distribution scheme*, by combining the expected locations of sensor nodes with the random subset assignment scheme in [42]. The resulting technique allows trade-offs between the security against node captures and the probability of establishing direct keys with a given memory constraint. Moreover, it does not require that the setup server be aware of the global network topology, making the deployment much easier.

We choose to improve the random subset assignment scheme using the expected locations because it can be considered as a generalization of several key pre-distribution schemes. Indeed, the basic probabilistic key pre-distribution scheme [20] is a special case of the random subset assignment scheme when each key is considered as a 0-degree polynomial [42]. Moreover, the key pre-distribution scheme in [18] is essentially equivalent to the random subset assignment scheme in [42]. Since most key pre-distribution schemes (except for the random pairwise keys scheme [12]) are based on the random and independent distribution of key units to sensor nodes, the results obtained through improving the random subset assignment scheme can be easily generalized to those schemes.

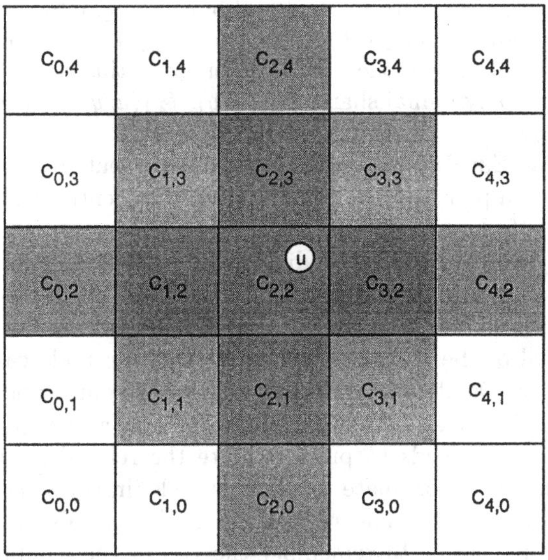

Fig. 4.4. Partition of a target field

The Closest Polynomials Pre-Distribution Scheme

Instead of randomly selecting polynomials for each sensor node as in the original random subset assignment scheme, the main idea of the proposed technique is to select polynomials for each sensor node based on the node's expected location. Specifically, we partition the target field into small areas called *cells* , each of which is associated with a unique random bivariate polynomial. Then, we distribute to each sensor node a set of polynomial shares that belong to the cells closest to the one that this sensor node is expected to locate in. For simplicity, we assume that the target field is a rectangle area that can be partitioned into equal-sized squares.

1. **Pre-Distribution:** The target field is first partitioned into a number of equal-sized squares $\{C_{i_c,i_r}\}_{i_c=0,1,...,C-1,i_r=0,1,...,R-1}$, each of which is a *cell* with the coordinate (i_c, i_r) denoting row i_r and column i_c. For convenience, we use $s = R \times C$ to denote the total number of cells. The setup server randomly generates s bivariate t-degree polynomials $\{f_{i_c,i_r}(x,y)\}_{i_c=0,1,...,C-1,i_r=0,1,...,R-1}$ and assigns $f_{i_c,i_r}(x,y)$ to cell C_{i_c,i_r}. Figure 4.4 shows an example partition of a target field.

   For each sensor node, the setup server first determines its *home cell* , in which this node is expected to locate. The setup server then discovers four cells adjacent to this node's home cell. Finally, the setup server distributes to the sensor node its home cell coordinate and the shares of the polynomials for its home cell and the four selected cells. For example, in Figure

4.4, node u is expected to be deployed in cell $C_{2,2}$. Obviously, cell $C_{2,2}$ is its home cell, and cells $C_{2,1}$, $C_{1,2}$, $C_{2,3}$ and $C_{3,2}$ are the four cells adjacent to its home cell. Thus, the setup server gives this node the coordinate $(2, 2)$ and the polynomial shares $f_{2,2}(u, y)$, $f_{2,1}(u, y)$, $f_{1,2}(u, y)$, $f_{2,3}(u, y)$, and $f_{3,2}(u, y)$.

2. **Direct Key Establishment:** After deployment, if two sensor nodes want to setup a pairwise key, they first need to identify a shared bivariate polynomial. If they can find at least one such polynomial, a common pairwise key can be established directly using the basic polynomial-based key pre-distribution. A simple way is to let the source node disclose its home cell coordinate to the destination node. From the coordinate of the home cell of the source node, the destination node can immediately determine the IDs of the polynomial shares the source node has.

3. **Sensor Addition and Revocation:** To add a new sensor node, the setup server only needs to pre-distribute the related polynomial shares and the home cell coordinate to the new node, in the same way as in the pre-distribution phase. The revocation method is also straightforward. Each node only needs to remember the IDs of the compromised sensor nodes that share at least one common bivariate polynomial with itself. Thus, in addition to the polynomial shares, the sensor node also needs to store a number of compromised sensor node IDs. If more than t nodes that share the same bivariate polynomial are compromised, a non-compromised sensor node that has a share of this polynomial simply removes the corresponding share and all the related compromised sensor node IDs from its memory.

Probability of Establishing Direct Keys

Similar to the analysis for the closest pairwise keys scheme, we also use $\dot{u}$ and $\bar{u}$ to represent the actual deployment location and the expected location of node u, respectively.

Consider two sensor nodes u and v. Since they are deployed independently, given the expected location of u and v, the conditional probability that they are neighbors can be calculated by

$$p(||\dot{u}, \dot{v}|| \leq d_r | \bar{u}, \bar{v}) = \int \int \int \int_{||\dot{u}, \dot{v}|| \leq d_r} \epsilon_{\bar{u}}(x_1, y_1) \epsilon_{\bar{v}}(x_2, y_2) \mathrm{d}x_1 \mathrm{d}y_1 \mathrm{d}x_2 \mathrm{d}y_2,$$

where $|| \cdot ||$ denotes the distance between two locations.

Assume these two sensor nodes u and v are expected to be deployed in cell C_{i_c, i_r} and C_{j_c, j_r}, respectively. To simplify our analysis, we assume that a sensor node is expected to locate randomly in its home cell. In other words, if sensor v is expected to be in cell C_{j_c, j_r}, then the probability density function for the expected location of v is $\frac{1}{L^2}$ for any location in the cell, and 0 otherwise. Therefore, the conditional probability that u and v are in each other's signal range given that u is expected to be deployed at location $\bar{u}$ and v is expected to be deployed in cell C_{j_c,j_r} can be calculated by

$$p(||\dot{u}, \dot{v}|| \leq d_r|\bar{u}, C_{j_c,j_r}) = \int\int_{C_{j_c,j_r}} \frac{p(||\dot{u}, \dot{v}|| \leq d_r|\bar{u}, (x, y))}{L^2} dxdy.$$

Hence, given u and v's home cells C_{i_c,i_r} and C_{j_c,j_r}, the probability of nodes u and v being able to directly communicate with each other can be estimated by

$$p(||\dot{u}, \dot{v}|| \leq d_r|C_{i_c,i_r}, C_{j_c,j_r}) = \int\int_{C_{i_c,i_r}} \frac{p(||\dot{u}, \dot{v}|| \leq d_r|(x, y), C_{j_c,j_r})}{L^2} dxdy.$$

Assume that on average, N_{cell} sensor nodes are expected to be deployed in each cell. Thus, among all the sensor nodes with home cell C_{j_c,j_r}, the average number of sensor nodes that the sensor node u with home cell C_{i_c,i_r} can directly communicate with can be estimated by $N_{cell} \times p(||\dot{u}, \dot{v}|| \leq d_r|C_{i_c,i_r}, C_{j_c,j_r})$. Therefore, overall, the average number of sensor nodes that u can directly communicate with can be estimated by

$$n_u = N_{cell} \cdot \sum_{\forall C_{j_c,j_r}} p(||\dot{u}, \dot{v}|| \leq d_r|C_{i_c,i_r}, C_{j_c,j_r}).$$

Let S_{i_c,i_r} denote the set of the home cells of the sensor nodes that share at least one common polynomial with the node whose home cell is C_{i_c,i_r}. According to the pre-distribution procedure, there are 13 such cells in each S_{i_c,i_r}. For example, Figure 4.4 shows $S_{2,2}$, which consists of all the shaded cells. Thus, the average number of neighbor sensor nodes that can establish a common key with u directly can be estimated by

$$n'_u = N_{cell} \cdot \sum_{C_{j_c,j_r} \in S_{i_c,i_r}} p(||\dot{u}, \dot{v}|| \leq d_r|C_{i_c,i_r}, C_{j_c,j_r}).$$

Hence, the probability of establishing a common key directly between u and a neighbor node of u can be estimated by

$$p = \frac{n'_u}{n_u} = \frac{\sum_{C_{j_c,j_r} \in S_{i_c,i_r}} p(||\dot{u}, \dot{v}|| \leq d_r|C_{i_c,i_r}, C_{j_c,j_r})}{\sum_{\forall C_{j_c,j_r}} p(||\dot{u}, \dot{v}|| \leq d_r|C_{i_c,i_r}, C_{j_c,j_r})}.$$

Similar to the analysis for the closest pairwise keys scheme, the above p can be used to estimate the probability of any node having a direct key with its neighbor node when the target field is an infinite field. For a limited field in our simulation, we simply use the probability p of the node with home cell $C_{C/2,R/2}$ having a direct key with its neighbor node to estimate the probability of having a direct key between any two neighbor nodes.

We use the simple deployment model described before to evaluate the performance, with signal range d_r as the basic unit for distance measurement

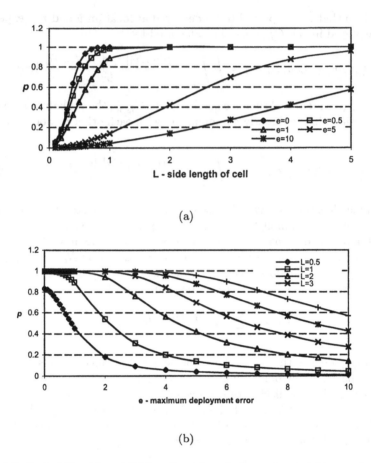

(a)

(b)

Fig. 4.5. Probability of establishing direct keys between two neighbor nodes given different cell side length L and maximum deployment error e

($d_r = 1$). Figure 4.5 shows the probability of establishing direct keys for different cell side length L and maximum deployment error e. Obviously, the probability of establishing direct keys increases as the cell side length L grows and decreases as the maximum deployment error e grows.

In general, the larger L is, the higher the probability of establishing a direct key between two neighbor nodes. However, the larger cell side length also leads to a larger number of sensor nodes sharing the same bivariate polynomial, which in turn degrades the security performance. Thus, we have to find the minimum value of L to meet the other constraints so that we can maximize the security performance. Figure 4.5 provides a guideline to determine the minimum value of L given the other constraints.

Security against Node Captures

According to the result of the polynomial-based key pre-distribution in [6], as long as no more than t polynomial shares of a bivariate polynomial are disclosed, an attacker knows nothing about the pairwise keys established through this polynomial between non-compromised nodes. Thus, the security of our scheme depends on the average number of sensor nodes sharing the same polynomial, which is equivalent to the number of sensor nodes that are expected to be located in a cell and its four adjacent cells.

As discussed in Section 4.1.2, the density of the sensor nodes in the network can be estimated by $D = \frac{m+1}{\pi d_r^2}$. The average number of sensor nodes that are expected to be located in a cell is $\frac{(m+1)L^2}{\pi d_r^2}$. Thus, the average number of sensor nodes that share the polynomial of a particular cell can be estimated by $N_s = \frac{5(m+1)L^2}{\pi d_r^2}$. Using the signal range as the basic unit of distance measurement $(d_r = 1)$, we have $N_s = \frac{5(m+1)L^2}{\pi}$.

We consider two types of attacks against the closest polynomials pre-distribution scheme. One is the *localized attack* , which targets the sensor nodes in a particular area in order to compromise the communication security in this area. The other is the *random attack* , which randomly selects sensor nodes to compromise.

In a localized attack, the attacker must compromise more than t out of N_s sensor nodes in order to compromise the direct keys between non-compromised sensor nodes in that area. In addition, the compromise of a particular area does not affect the direct keys in any other area because all bivariate polynomials are chosen randomly and independently.

Consider a random attack. We assume a fraction p_c of sensor nodes in the network have been compromised by an attacker. This means that each sensor node has the probability of p_c being compromised. Thus, among N_s sensor nodes that have polynomial shares of a particular cell, the probability that exactly i sensor nodes have been compromised can be estimated by

$$P_c(i) = \frac{N_s!}{(N_s - i)!i!}p_c^i(1 - p_c)^{N_s-i}.$$

Therefore, the probability that the bivariate polynomial assigned to this cell is compromised, which is equivalent to the probability of a direct key between two non-compromised nodes being compromised, can be estimated by $P_c = 1 - \sum_{i=0}^{t} P_c(i)$. Figure 4.6 includes the relationship between the fraction of compromised direct keys for non-compromised sensor nodes and the fraction of compromised nodes under different combinations of m and L given the storage capacity that is equivalent to 200 cryptographic keys $(t = 39)$. An interesting result is that regardless of the total number of sensor nodes in the network, the less the density of the sensor network, the higher the security guarantee it can provide.

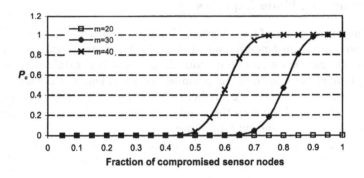

(a) L=1

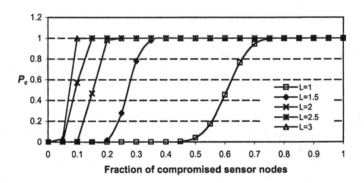

(b) m=40

Fig. 4.6. Fraction of compromised direct keys between non-compromised sensor nodes v.s. fraction of compromised sensor nodes. Assume each node has available storage equivalent to 200 cryptographic keys.

Overhead

In this scheme, each sensor node needs to store the coordinate of its home cell and the polynomial shares of five cells. The storage overhead for the coordinate of its home cell is negligible. Thus, each sensor node needs to allocate $5(t+1)\log q$ memory space to store the secret. When there are compromised sensor nodes, each non-compromised sensor node also needs to store the IDs of the compromised sensor nodes with which it shares at least one common polynomial. However, for each of the 5 polynomials, a non-compromised sensor node only needs to store up to t IDs; it can remove the corresponding polynomial share and all the related IDs if the number of compromised sensor nodes sharing the polynomial exceeds t.

To establish a common key between two neighbor nodes, one of them initiates a request by disclosing its home cell coordinate. Once the other party receives such a message, it can immediately determine the common pairwise key and reply a message to identify the corresponding key. Thus, the communication overhead includes two short messages.

To compute the common key with a given sensor node, each sensor node needs to evaluate a t-degree polynomial. Thus, the computation overhead in each sensor node mainly comes from the evaluation of this polynomial, which can be done efficiently by using the optimization technique in [42].

Improvements

Compared with the original random subset assignment scheme in [42], the closest polynomials pre-distribution scheme can achieve better performance due to the explicit usage of expected locations. First, given certain storage constraint and the required probability of sharing direct keys between sensor nodes, the random subset assignment can only tolerate a small number of compromised sensor nodes, while the closest polynomials pre-distribution scheme can tolerate a large fraction of compromised nodes. Figure 4.7 shows that the security can be improved significantly by using prior deployment knowledge of sensor nodes. (To save space, this figure also includes the security performance of other techniques, which will be discussed in the later comparison.) Second, the probability of sharing direct keys between sensor nodes in the random subset assignment scheme is fixed given certain polynomial pool size and the storage constraint on sensor nodes; while for the closest polynomials pre-distribution scheme, this probability is independent from the total number of polynomials in the pool. Indeed, it can be further improved by increasing cell side length L for a given maximum deployment error e as shown in Figure 4.5.

Comparison

Now let us compare our scheme in this subsection with the previous methods (the basic probabilistic scheme [20], the q-composite scheme [12], the random pairwise keys scheme [12], the grid-based scheme [42], and the closest pairwise keys scheme). Evaluation of those schemes requires the network size. To be fair, we use the following method to estimate the network size. Assume that on average, there are m sensor nodes that fall into each sensor's signal range. Based on the analysis in [12], we estimate the total number of sensor nodes in the network is $N = 2^{mp}$ to make sure the network is fully connected with a high probability if the node only contacts its neighbor nodes, where p is the probability of establishing a direct key between two neighbor sensor nodes.

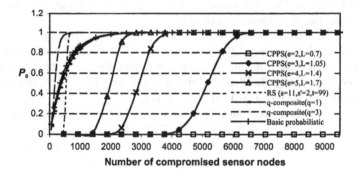

Fig. 4.7. Fraction of compromised direct keys between non-compromised sensor nodes v.s. number of compromised sensor nodes. Assume each node has available storage equivalent to 200 cryptographic keys. Assume $p = 0.33$ and $m = 40$. CPPS denotes the closest polynomials pre-distribution random scheme.

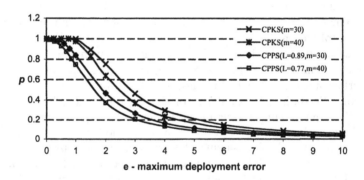

Fig. 4.8. Probability of establishing pairwise key directly between two neighbor nodes given different e and m. The length of cell side in CPPS is configured so that it is perfectly resistant to the node captures. Assume each node has available storage equivalent to 200 cryptographic keys.

Let us first compare our new scheme with the basic probabilistic scheme [20] and the q-composite scheme [12]. Figure 4.7 compares the fraction of compromised direct keys shared between non-compromised sensor nodes given the same p, m, and storage overhead. We can see that our scheme performs significantly better than the other two schemes. It also shows that the more precise the sensor deployment is, the higher security it can guarantee.

We then compare our new scheme in this subsection with the random pairwise keys scheme [12]. By limiting the number of sensor nodes sharing the same bivariate polynomial, our proposed scheme can be modified to provide perfect security against node captures. In this case, we have $N_s = \frac{5(m+1)L^2}{\pi} \le (t+1)$. From the previous result, we know that the value of p only depends on L and e. Thus, given the same probability of establishing direct keys between sensor nodes, our proposed scheme has no limit on the total number of sensor nodes it can support. However, the random pairwise key scheme can only support at most $\frac{c}{p}$ sensor nodes, where c is the number of cryptographic keys a sensor node stores [12]. Thus, our new scheme can achieve better performance when the expected location information is available.

Similar to the closest pairwise keys scheme, the closest polynomials pre-distribution scheme has some advantages over the grid-based scheme in [42]. First, in order to provide certain security guarantee against node compromise attacks, the grid-based scheme has limitations on the maximum supported network size given certain storage constraint as pointed out in [42]. However, the closest polynomials pre-distribution scheme has no limitations on the maximum supported network size given reasonable maximum deployment errors. Second, the closest polynomials pre-distribution scheme provides higher probability of establishing direct keys between sensor nodes than the grid-based scheme. Though the grid-based scheme can guarantee to establish a direct or indirect key between any two sensor nodes, it requires that any two sensor nodes can communicate with each other, which may not be true in real applications.

Now we compare our new scheme proposed in this subsection with the closest pairwise keys scheme in Section 4.1.2. For the closest pairwise key pre-distribution, given a fixed storage capacity c, signal range d_r, node density D, and the maximum deployment error e, the probability of establishing direct keys between sensor nodes is fixed. However, for our new scheme proposed in this subsection, given the above constraints, it can still achieve arbitrary high probability to establish direct keys between sensor nodes by increasing the cell side length L as shown in Figure 4.5. For example, in the closest pairwise keys scheme, if $\gamma = 5$, $e = 3$, the probability of having a common pairwise key between two neighbor nodes is 0.4. In contrast, our new scheme allows us to increase cell side length to achieve a higher probability of establishing direct keys between neighbor sensor nodes and still provide certain degree of security.

An advantage of the closest pairwise key scheme is that the compromise of sensor nodes does not lead to the compromise of direct keys shared between non-compromised sensor nodes. By having $N_s \le (t+1)$, the closest polynomials pre-distribution scheme can also provide this security property. To further compare these two schemes under this condition, Figure 4.8 shows the probabilities of establishing direct keys under different node densities and maximum deployment errors, assuming the storage capacity is equivalent to 200 cryptographic keys. We can see that although the closest pairwise keys scheme has a higher probability to establish direct keys between neighbor sensor nodes, our new scheme is not significantly worse. Considering the flexibility to trade-off security and performance in the closest polynomials pre-distribution scheme, we conclude that this scheme is more desirable than the closest pairwise keys scheme in certain applications.

4.2 Improving Key Pre-Distribution with Post Deployment Knowledge

In this section, we propose to take advantage of the post deployment knowledge of sensor nodes to improve the pairwise key pre-distribution in static sensor networks. The main idea is to assign each sensor node an excessive amount of pre-distributed keys by using the memory for sensing applications, prioritize the pre-distributed keys based on post deployment knowledge, and discard low priority keys to thwart node compromise attacks and return memory to the applications. We call this process *key prioritization* .

We do not assume any prior knowledge of sensors' locations. However, we assume that every sensor node can discover its real deployment location securely after the deployment of sensor networks. This assumption is practical. As pointed out in [1], "most of the sensing tasks require the knowledge of positions," and "location finding systems are also required by many of the proposed sensor network routing protocols." Indeed, there have been a series of recent advances in determining individual sensor nodes' positions (with a global positioning system (GPS) or local references) [40, 55], as well as securing location discovery [70, 39, 19, 46, 47]. Thus, we believe that in many sensor network applications, it is possible for the sensor nodes to determine their deployment locations securely.

Using memory for applications to store an excessive amount of pre-distributed keys is practical in sensor networks. Though sensor nodes are memory constrained, EEPROM, which is usually used to save sensed data, is much more plentiful than RAM on a sensor node. For example, a typical MICA2 mote [14] comes with 512KB EEPROM, but only 4KB RAM. Thus, we may store an excessive amount of keying materials. However, in this situation, the compromise of a sensor node reveals more secrets in the network. To deal with this problem, we propose to remove the keying materials that are less likely to be used (based on the post deployment knowledge). We further assume that an attacker cannot recover the removed keys at sensor nodes even if these nodes are compromised later. Moreover, the removal of low priority keys also returns memory to sensing applications, which may be desirable in certain scenarios.

Note that accessing EEPROM is more expensive than accessing RAM in a typical sensor node. It takes more energy to delete cryptographic keys from the EEPROM than to delete keys in RAM. Based on the results in [74], the energy consumed by writing 16 bytes to EEPROM is close to the energy consumed by computing for 237,360 clock cycles (about 29.67 ms). Nevertheless, the key prioritization technique only requires such deletion operations once for each node during the entire lifetime. Therefore, we believe that such deletion operations are feasible in the current generation of sensor networks.

For the sake of presentation, we refer to the pre-distributed keying materials used in a key pre-distribution scheme (e.g., [20, 12, 18, 42, 81, 43]) as key units. More specifically, a *key unit* is a minimal piece of keying material from which a valid key can be derived. A key unit in the probabilistic key pre-distribution scheme [20], the q-composite scheme [12], or the random pairwise keys scheme [12] is simply a pre-distributed key. In the polynomial pool-based key pre-distribution schemes [42], a key unit is a t-degree polynomial from which a node can compute keys shared with others. In the pairwise key pre-distribution scheme presented in [18], a key unit is a row of the secret matrix A_i in a key space S_i. A common property of the key units in all these schemes is that *two sensor nodes sharing the same or relevant key units can derive a common key.*

In the following, we first present an approach of key prioritization in static sensor networks and then show how to improve the random subset assignment scheme [42] using this approach.

4.2.1 Key Prioritization Using Post Deployment Knowledge

By using memory for sensing applications, a sensor node can keep a large number of key units during key pre-distribution. By prioritizing these key units based on post deployment knowledge, a sensor node can give up the key units that are less likely to be used for pairwise key establishment to thwart node compromise attacks and return the memory to the sensing applications. As a result, it has a higher probability to keep those key units that may be required for secure communications with its neighbor nodes.

Specifically, we prioritize pre-distributed key units based on the deployment locations of sensor nodes. In order to do so, we map each key unit to a location in the sensor network field before deployment. After the sensor network is deployed, if a sensor node can discover its location, it can prioritize the pre-distributed key units based on this location. The node may rank all the key units according to the distances between its location and the locations of key units so that the closer a key unit is to the sensor node, the higher priority it has. As a result, sensor nodes close to each other are more likely to keep a common key unit than those that are far away from each other, and thus have a higher probability to establish a common key.

An attractive feature of using deployment locations is that there is almost no overhead. Once determining its location, a node only needs to perform simple computation to rank the pre-distributed key units, and no communication with other sensor nodes is required. Moreover, this approach allows incremental deployment of sensor nodes since the only information a sensor node needs to prioritize its key units is its own location.

The proposed approach can be used to improve many key pre-distribution techniques (e.g. the basic probabilistic scheme [20], the q-composite scheme [12], the random subset assignment scheme in [42]). However, the random pairwise keys scheme is based on a different approach [12], where each key is related to two particular sensor nodes. This makes it useless to apply the above approach in such a scheme. Nevertheless, the random pairwise keys scheme can still have a better performance by loading an excessive amount of keys in the memory for sensing applications. Since it provides the perfect security guarantee, it is unnecessary to thwart node compromise attacks by removing low priority keys from memory when the applications have enough memory.

4.2.2 Improving Random Subset Assignment Scheme with Deployment Locations

In the random subset assignment scheme, the more polynomial shares a node has, the more likely it can establish a common key with other sensor nodes. The improved scheme reuses the memory for sensing applications to keep more polynomial shares during key pre-distribution, gives higher priority to the polynomial shares that are most likely to be used after the deployment location is known, and discards low-priority polynomial shares to thwart node compromise attacks and returns memory to the applications.

The Improved Scheme

The details of the improved scheme are described below. The improvements are mainly in the key pre-distribution and key prioritization phases.

1. **Key Pre-Distribution:** The key pre-distribution phase consists of two stages. In the first stage, the setup server randomly generates a set $\mathcal{F}$ of bivariate t-degree polynomials and associates each polynomial with a unique location in the target field. These locations are evenly distributed over the entire target field. For the sake of presentation, we use $f_{x,y}$ to denote the bivariate polynomial in $\mathcal{F}$ associated with the location coordinate (x, y). For convenience, we also use the location coordinates as the IDs of the corresponding bivariate polynomials. In the second stage, for each sensor node, the setup server randomly picks a set of c bivariate polynomials from the polynomial pool and distributes the corresponding polynomial shares as well as their locations to the sensor node.

2. **Key Prioritization:** After deployment, each sensor node first determines its location. Then based on this location and the locations associated with the pre-distributed polynomial shares, the sensor node ranks

the polynomial shares in terms of their distances to its deployment location. The closer a polynomial share to this sensor node, the higher priority it has. For simplicity, we assume the sensor node chooses the highest c' polynomial shares to save, where c' is the number of shares it keeps in the memory reserved for keys. (A sensor node may keep more polynomial shares until the sensing applications require the corresponding memory. However, as we will show in our analysis, this makes the sensor network more vulnerable to node compromises.)

3. **Direct Key Establishment:** This phase can be performed in the same way as in the original random subset assignment scheme [42]. To establish a direct key between two sensor nodes, they only need to identify a common bivariate polynomial shared between them, which can be achieved by exchanging the IDs of polynomial shares that they have.

4. **Addition and Revocation:** To add a new sensor node, the setup server only needs to perform the above key pre-distribution and key prioritization in the same way. The revocation method is also straightforward. Each node only needs to remember the IDs of the compromised sensor nodes that share at least one common bivariate polynomial with itself. If more than t nodes that share the same bivariate polynomial are compromised, a non-compromised sensor node that has a share of this polynomial simply removes the corresponding share and all the related compromised sensor node IDs from its memory.

Probability of Establishing Direct Keys between Neighbor Nodes

The ideas of using prior deployment and post deployment knowledge to improve the performance of key pre-distribution are two independent techniques. They can combine together easily to achieve better performance and security. In the rest of this subsection, we only analyze the improvements on the performance and security introduced by using key prioritization based on post deployment knowledge.

Similarly, we use the signal range d_r as the basic unit to measure distances ($d_r = 1$). Assume each sensor node has m neighbor sensor nodes on average. In other words, there are $(m+1)$ sensor nodes on average in an area of $\pi \cdot d_r^2 = \pi$. We further assume the size of the target field is S. Then the total number of sensor nodes in the network can be estimated as $\frac{S \times (m+1)}{\pi}$.

Consider a sensor node u that gets a set of c polynomial shares during the pre-distribution phase and keeps c' of them after key prioritization. Based on the key prioritization process, node u keeps c' polynomial shares whose associated locations are no more than r away from u's location, where $r = \sqrt{\frac{S \times c'}{\pi c}} = \sqrt{\frac{c' \times N}{c \times (m+1)}}$. In other words, all the bivariate polynomial shares that fall into the circle centered at this node's location with radius r are still in the sensor node's memory. For convenience, we call such a circle the *share circle*. The left part of Figure 4.9 illustrates the share circle and the signal range of

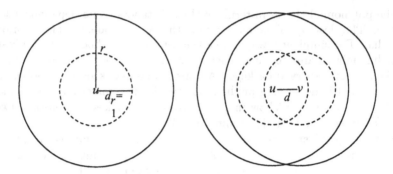

Fig. 4.9. Shared circles of neighbor sensor nodes

node u. The inner solid circle represents the area in which u can communicate with other sensor nodes directly, while the outer dashed circle represents the area in which the c' polynomial shares that u keeps fall.

Consider a pair of neighbor sensor nodes u and v. As illustrated in the right part of Figure 4.9, the polynomials that u and v share fall into the outer circles of both u and v. In the following, we first estimate how many polynomial shares u or v has in this area, and then estimate the probability that u and v have shares of a common polynomial (i.e., the probability that they can establish a direct key).

Assume the distance between u and v is d. Since they are neighbors, we have $d < d_r = 1$. It is easy to see that in a large sensor network, the total number of sensor nodes in the network is usually much larger than the number of neighbor nodes for a particular sensor node. Thus, we usually have $c' \times N \geq c \times (m+1)$, which implies $r \geq 1$ (since $r = \sqrt{\frac{c' \times N}{c \times (m+1)}}$). Then we can estimate the size of the overlapped area of the share circles of u and v as

$$S_o(d) = 2 \times \frac{2 \times \arccos(d/2r)}{2\pi} \times \pi \times r^2 - d \times \sqrt{r^2 - d^2/4}.$$

On average, the number of pre-distributed polynomial shares that fall into this overlapped area for both u and v can be estimated by $n(d) = \lfloor c \times \frac{S_o(d)}{S} \rfloor$. In other words, both u and v have shares of about $n(d)$ polynomials that fall into this area. We can further estimate the total number of polynomials that are distributed over this area as $n_t(d) = \lfloor |\mathcal{F}| \times \frac{S_o(d)}{S} \rfloor$. Therefore, the probability that sensor nodes u and v share a common polynomial can be estimated by

$$p(d) = 1 - \prod_{i=0}^{n(d)-1} \frac{n_t(d) - n(d) - i}{n_t(d) - i}.$$

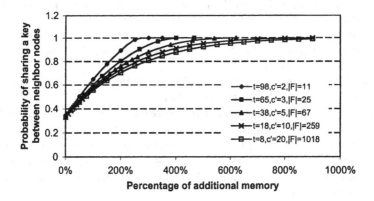

Fig. 4.10. Probability of sharing a direct key between neighbor sensor nodes.

Thus, the (average) probability of sharing a polynomial between neighbor sensor nodes, which is equivalent to the probability of establishing a common direct key between two neighbor sensor nodes, can be estimated by

$$p = \int_0^{d_r} \int_0^{2\pi} \frac{p(\rho)\rho}{\pi d_r^2} \mathrm{d}\rho \mathrm{d}\theta = 2 \int_0^1 p(\rho)\rho \mathrm{d}\rho.$$

Figure 4.10 shows the probabilities of establishing direct keys between neighbor sensor nodes using key prioritization with different parameters. We assume the sensor network has $N = 10,000$ sensor nodes, the average number of neighbor sensor nodes is $m = 30$, and each sensor node's memory for key units after key prioritization (c') is equivalent to 200 cryptographic keys. The number of polynomials in the polynomial pool $\mathcal{F}$ is chosen to make the probability of sharing a direct key between two neighbors be 0.33 if no additional memory is allocated to store polynomial shares in the key pre-distribution phase. We can clearly see that the probability of sharing a direct key between two neighbor sensor nodes is improved significantly as the memory (c) allocated for pre-distributed polynomial shares (before key prioritization) increases.

Security Analysis

Note that there are more polynomial shares stored in a sensor node before key prioritization than after it. Thus, compromising sensor nodes before the key prioritization phase reveals more secrets than compromising the same set of sensor nodes after it. An attacker may take advantage of this observation and attack the network between the key pre-distribution and key prioritization phases. However, once a sensor node determines its location, it can finish key prioritization instantly. For convenience, we refer to the time period between pre-distributing key units to a sensor node and completing key prioritization in the sensor node as the *window of vulnerability* . Intuitively, the shorter the window of vulnerability is, the fewer secrets may be disclosed due to compromised sensor nodes.

In the following, we evaluate the ability of the improved scheme to tolerate compromised sensor nodes in two situations.

Situation 1: No node compromises during the window of vulnerability. In this situation, we assume the sensor network is well-protected during the window of vulnerability. That is, the sensor network is assumed to be secure and have none of its sensor nodes compromised between key pre-distribution and key prioritization. After this time period, the network may be exposed to attacks.

Assume an attacker randomly compromises N_c sensor nodes in the network after the window of vulnerability. Consider any polynomial f in $\mathcal{F}$. The probability that a sensor node has a polynomial share of f is $\frac{c'}{|\mathcal{F}|}$. Then we can estimate the probability that exactly i out of N_c compromised sensor nodes have shares of this polynomial by

$$P_c(i) = \frac{N_c!}{(N_c - i)!i!}(\frac{c'}{|\mathcal{F}|})^i(1 - \frac{c'}{|\mathcal{F}|})^{N_c-i}.$$

According to the results in [6], an attacker can compute any key generated using a t-degree polynomial if he/she has at least $t + 1$ distinctive shares of this polynomial. Thus, the probability of f being compromised is $P_c = 1 - \sum_{i=0}^{t} P_c(i)$. Since f is any polynomial in $\mathcal{F}$, the fraction of compromised direct keys between non-compromised sensor nodes can be estimated as P_c. The ratio $\frac{c'}{|\mathcal{F}|}$ in this formula also implies that given the same value of t, the security performance against random node compromises will be the same if the ratio between the number of shares stored in one sensor node after key prioritization and the total number of polynomials in the polynomial pool is the same. Let us revisit Figure 4.10. Every line in this figure has the same ratio $\frac{c'}{|\mathcal{F}|}$ (and thus the same security performance against random node compromises). The shape of each line indicates that by using more extra memory for key units before key prioritization, we can significantly improve the probability of sharing common keys between neighbor sensor nodes without reducing the security.

Figure 4.11 shows the security performances of the improved scheme under different conditions. Following [42], we evaluate the security performance using the fraction of compromised direct keys between non-compromised sensor nodes given a number of compromised sensor nodes. We assume each sensor node keeps key materials equivalent to 200 keys after key prioritization, the probability of sharing a common key between two sensor nodes is 0.33, and each sensor node only keeps $c' = 2$ key units after key prioritization. Parameter c represents the number of key units distributed to each sensor node during key pre-distribution. In the special case of $c = 2$, the improved scheme becomes the original random subset assignment scheme proposed in [42]. (Note

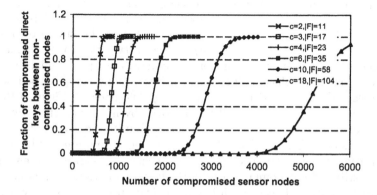

Fig. 4.11. Security performance of the improved scheme in Situation 1.

that for different values of c, we need a different number of polynomials in the polynomial pool to maintain the same probability of sharing common keys between neighbor sensor nodes.) Figure 4.11 shows that when c increases by 1, the vertical line shifts to the right significantly. This means the security is improved significantly as the additional memory for polynomial shares at the pre-distribution stage increases.

Situation 2: Limited node compromises during the window of vulnerability. In the second situation, we assume an attacker is able to compromise up to N_t sensor nodes after key pre-distribution but before key prioritization. We consider the worst case in the following analysis; that is, the attacker has compromised N_t sensor nodes. Assume the attacker randomly compromises N_c sensor nodes after the key prioritization phase. Consider any polynomial f in $\mathcal{F}$. The probability that a sensor node compromised before key prioritization has a polynomial share of f is $\frac{c}{|\mathcal{F}|}$. Similarly, the probability that a sensor node compromised after key prioritization has a polynomial share of f is $\frac{c'}{|\mathcal{F}|}$. Thus, the probability that exactly i compromised sensor nodes have polynomial shares of f can be calculated by

$$P_c(i) = \sum_{j+k=i} \frac{N_t!}{(N_t-j)!j!} (\frac{c}{|\mathcal{F}|})^j (1-\frac{c}{|\mathcal{F}|})^{N_t-j} \frac{N_c!}{(N_c-k)!k!} (\frac{c'}{|\mathcal{F}|})^k (1-\frac{c'}{|\mathcal{F}|})^{N_t-k}.$$

Therefore, the probability of this polynomial being compromised can be estimated as $P_c = 1 - \sum_{i=0}^{t} P_c(i)$. Since f is any polynomial in the polynomial pool, the fraction of compromised direct keys between non-compromised sensor nodes can be also estimated as P_c.

Figure 4.12 shows the security performance of the improved scheme when an attacker compromises a few sensor nodes before the sensor nodes finish key prioritization. Similar to Situation 1, we assume each sensor node keeps key materials equivalent to 200 keys after key prioritization, the probability of sharing a common key between two sensor nodes is 0.33, and each sensor node only keeps $c' = 2$ key units after key prioritization. When there are only a small number of compromised sensor nodes before the key prioritization phase (e.g., 100 nodes in Figure 4.12(a)), the security performance is enhanced significantly by increasing the number of pre-distributed polynomial shares at the pre-distribution phase. When there are too many compromised sensor nodes before the key prioritization phase (e.g., 500 nodes in Figure 4.12(b)), the security performance can still be improved by increasing the number of polynomial shares at the pre-distribution phase, but it is not as significant as in the previous case. In general, the fewer sensor nodes compromised before key prioritization, the more improvement we can achieve by increasing the number of pre-distributed polynomial shares in the key pre-distribution phase.

Since a sensor node can finish key prioritization almost instantly after it determines its location, we believe the sensor nodes can be protected fairly well during the window of vulnerability. Indeed, a sensor node is not vulnerable at the same level during the entire window of vulnerability. The most vulnerable period of time is the period after deployment and before key prioritization, which is actually quite short due to the ease of completing key prioritization. Thus, we believe that it is unlikely that an attacker can compromise a large number of sensor nodes before they finish key prioritization.

Overheads

Since the improved scheme uses application memory only before key prioritization, it has almost the same overheads as the original scheme. Specifically, each sensor node has to store c polynomial shares and the locations associated with these shares in the key pre-distribution phase, and only keeps c' of them after determining its deployment location. A location coordinate can be represented with integers much smaller than a polynomial share (e.g., 4 bytes to encode x or y coordinate). Each location coordinate takes approximately the same space as one key. Thus, the storage overhead in a sensor node is approximately $c(t + 2) \log q$ in the key pre-distribution phase. Similarly, after key prioritization, the storage overhead for key units is about $c'(t + 2) \log q$. In addition, each sensor node needs to remember the IDs of compromised sensor nodes with which this node shares at least one common polynomial. This introduces at most $c't \log q'$ storage overhead after key prioritization, where q' is a number that is large enough to accommodate all sensor nodes in the network. To establish a common key with a given neighbor node, two sensor nodes only need to exchange their polynomial IDs. To compute a pairwise key, both sensor nodes need to evaluate a t-degree polynomial, which can be done efficiently with the optimization technique in [42].

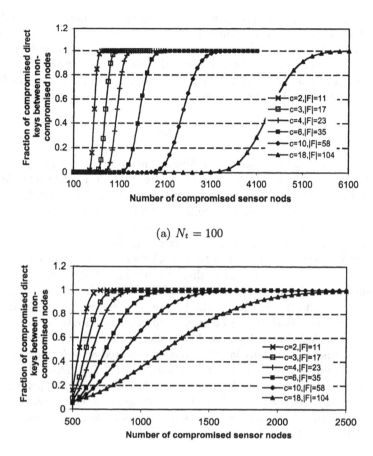

(a) $N_t = 100$

(b) $N_t = 500$

Fig. 4.12. Security performance of the improved scheme in Situation 2.

4.3 Improving Key Pre-Distribution with Group Deployment Knowledge

In this section, we introduce a practical deployment model, where sensor nodes are only required to be deployed in groups. The knowledge used to improve the performance of key pre-distribution is the assumption that the sensor nodes belonging to the same group are deployed close to each other. This assumption is generally true since the sensor nodes in the same group are supposed to be deployed from the same point at the same time. For example, a group of sensor nodes are dropped from the helicopter during the deployment. For the sake of presentation, we call such a group of sensor nodes a *deployment group* .

We assume that sensor nodes are static once they are deployed. We define the *resident point* of a sensor node as the point location where this sensor node finally resides. Sensors' resident points are generally different from each other. However, we assume the resident points of the sensor nodes in the same group follow the same probability distribution function. The detailed description of the deployment model is given below.

4.3.1 Group-Based Deployment Model

The sensor nodes to be deployed are divided into n groups $\{G_i\}_{i=1,...,n}$. The nodes in the same deployment group G_i are deployed from the same place at the same time with the deployment index i. During the deployment, the resident point of any node in group G_i follows a probability distribution function $f_i(x, y)$, which we call the *deployment distribution* of group G_i. An example of the pdf $f_i(x, y)$ is a two-dimensional Gaussian distribution.

The actual deployment distribution is affected by many factors. For simplicity, we model the deployment distribution as a Gaussian distribution (also called Normal distribution). Gaussian distribution is widely studied and proved to be useful in practice. Although we only employ the Gaussian distribution, our methodology can be applied to other distributions as well.

We assume that the deployment distribution for any node in group G_i follows a two-dimensional Gaussian distribution centered at a *deployment point* (x_i, y_i). *Different from the deployment models in [17, 30], where the deployment points of groups are pre-determined, we do not assume any prior knowledge of such deployment points.* In fact, we only assume the existence of such deployment points. The mean of the Gaussian distribution μ equals (x_i, y_i), and the pdf for any node in group G_i is the following:

$$f_i(x, y) = \frac{1}{2\pi\sigma^2} e^{-[(x-x_i)^2 + (y-y_i)^2]/2\sigma^2} = f(x - x_i, y - y_i),$$

where σ is the standard deviation, and $f(x, y) = \frac{1}{2\pi\sigma^2} e^{-(x^2+y^2)/2\sigma^2}$.

According to the deployment model discussed before, the sensor nodes in the same deployment group have high probability of being neighbors. To take advantage of this observation, the pairwise key pre-distribution techniques should at least benefit the sensor nodes in the same deployment group. Hence, we first employ an *in-group key pre-distribution* method, which enables the sensor nodes in the same deployment group to establish pairwise keys between each other with high probability. To handle the pairwise key establishment between sensor nodes in different deployment groups, we then employ a *cross-group key pre-distribution* method, which enables selected sensor nodes in different deployment groups to establish pairwise keys and thus bridges different deployment groups together.

In the above idea, as long as a key pre-distribution technique can provide pairwise key establishment between sensor nodes in a group, it can be used as the basic building block to construct the group-based scheme. This implies that our framework can be applied to any existing key pre-distribution technique.

4.3.2 A General Framework

Without loss of generality, let $\mathcal{D}$ denote the key pre-distribution technique used in the framework. This subsection shows how to construct an improved key pre-distribution technique by applying the group knowledge to $\mathcal{D}$.

A key pre-distribution technique can usually be divided into three phases, *pre-distribution*, which specifies how to pre-distribute keying materials to each sensor node, *direct key establishment*, which specifies how to establish a pairwise key shared between two sensor nodes *directly*, and *path key establishment*, which specifies how to find a sequence of nodes to help two given nodes establish a temporary session key. The key established in the direct key establishment phase is called the *direct key*, while the key established in the path key establishment phase is called the *indirect key*.

We refer to an instantiation of $\mathcal{D}$ for a group of sensor nodes as a *key pre-distribution instance* . A key pre-distribution instance D includes a set of target sensor nodes G, a set of keying materials K (e.g., keys [20, 12], polynomials [42], or matrixes [18]), and a function g that maps an ID in G to a subset of keying materials in K. In such an instance, each sensor node i in group G is pre-distributed with a set of secrets that are computed from the mapping result of ID i under function g. This set of secrets could be keys [20, 12], polynomial shares [42], or a row of elements on a matrix [18].

We also define the following *property functions* to characterize the typical properties of a key pre-distribution instance.

- $M(D)$: the memory requirements on sensor nodes for a key pre-distribution instance D.
- $p_{dk}(D)$: the probability of sharing a direct key between any two sensor nodes in a key pre-distribution instance D.
- $p_{cd}(D, x)$: the probability of a direct key between two non-compromised sensor nodes being compromised in a key pre-distribution instance D when the adversary has randomly compromised x sensor nodes.

Our group-based framework is built upon a number of key pre-distribution instances. For simplicity, we assume there are n equal size deployment groups with m sensor nodes in each of those groups. The details of the our framework are described below.

Pre-Distribution

For each deployment group G_i, we randomly generate a key pre-distribution instance D_i. The pairwise key establishment between sensor nodes in group

G_i is based on instance D_i. For the sake of presentation, these randomly generated instances are called the *in-group (key pre-distribution) instances*.

To handle the pairwise key establishment between sensor nodes in different deployment groups, we further generate m key pre-distribution instances $\{D'_i\}_{i=1,...,n}$. These instances are called the *cross-group (key pre-distribution) instances*. The set of sensor nodes having the same cross-group instance D'_i form a *cross group* G'_i. The requirements on these cross groups $\{G'_1, ..., G'_m\}$ are: (1) each cross group includes exactly one sensor node from each deployment group, and (2) there are no common sensor nodes between any two different cross groups. In other words, for any i and j with $i \neq j$, we have $G'_i \cap G'_j = \phi$ and $|G'_i \cap G_j| = 1$. By doing this, each cross group provides a potential link for any two deployment groups.

In the following, we propose a simple way to construct deployment groups and cross groups. Basically, each deployment group G_i contains the sensor nodes with IDs $\{(i-1)m+j\}_{j=1,...,m}$, while each cross group G'_i contains the sensor nodes with IDs $\{i+(j-1)m\}_{j=1,...,n}$. Figure 4.13 shows an example of such a construction when $n = 4$ and $m = 3$. In the figure, G'_1 includes nodes 1, 4, 7 and 10, G'_2 includes nodes 2, 5, 8 and 11, and G'_3 includes nodes 3, 6, 9 and 12.

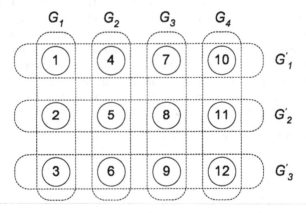

Fig. 4.13. Example of group construction

Direct Key Establishment

After pre-distribution, each sensor node belongs to two key pre-distribution instances, an in-group instance and a cross-group instance. Hence, the direct key establishment between two sensor nodes is simple and direct. If they are in the same deployment group, for example G_i, they can follow the direct key establishment of the in-group instance D_i. If they are not in the same deployment group but belong to the same cross group G'_j, they can follow the direct key establishment of the cross-group instance D'_j. To determine if two sensor nodes are in the same deployment group or the same cross group, they only need to exchange the IDs of groups that they belong to. In our framework, they only need to know the ID of the other party due to our group construction method.

Path Key Establishment

If two nodes cannot establish a direct key, they have to go through path key establishment to find a number of other sensor nodes to help them establish an indirect key. Similar to the direct key establishment, if two nodes are in the same deployment group G_i, they can follow the path key establishment in D_i. The indirect keys between sensor nodes in the same group are called the *in-group indirect keys*. When two nodes belong to two different groups G_i and G_j, we use a different method to establish an indirect key. Basically, we need to find a "bridge" between these two deployment groups in order to setup a *cross-group indirect key*. A bridge between group G_i and G_j is defined as a pair of sensor nodes $\langle a, b \rangle$ ($a \in G_i$ and $b \in G_j$) that belong to the same cross group G'_k ($a, b \in G'_k$). A bridge is valid when the two sensor nodes involved in this bridge can establish a direct key.

According to the pre-distribution step, there are m potential bridges (one from each cross group) that can be used to establish an indirect key. In addition, due to our group construction method, a sensor node can easily compute all possible bridges between any two deployment groups. Specifically, the possible bridges between group G_i and G_j are $\{\langle (i-1)m+k, (j-1)m+k \rangle\}_{k=1,\ldots,m}$. For example, there are 3 bridges between group G_1 and G_4 in Figure 4.13: $\langle 1, 10 \rangle$, $\langle 2, 11 \rangle$, and $\langle 3, 12 \rangle$.

Assume every message between two sensor nodes is encrypted and authenticated by the pairwise key established between them. The path key establishment for the sensor nodes in different deployment groups works as follows.

1. The source node u first tries the bridge involving itself to establish an indirect key with the destination node v. Assume this bridge is $\langle u, v' \rangle$. Node u first sends a request to v' if it can establish a direct key with v'. If node v' can also establish a (direct or indirect) key with the destination node v, node v' forwards this request to the destination node v to establish an indirect key.
2. If the first step fails, node u tries the bridge that involves the destination node v. Assume the bridge is $\langle u', v \rangle$. In this case, node u sends a request to node u' if it can establish a (direct or indirect) key with u'. If node u' can establish a direct key with node v, it forwards the request to the destination node v to establish an indirect key. Note that if node u and v are in the same cross group, this step can be skipped since step 1 and step 2 compute the same bridge.

3. When both of the above steps fail, node u has to try other bridges. Basically, it randomly chooses a bridge $\langle u', v' \rangle$ other than the above two, assuming u' is in the same deployment group with u, and v' is in the same deployment group with v. Node u then sends a request to u' if it can establish a (direct or indirect) key with u'. Once u' receives this request, it forwards the request to v' in the bridge if they share a direct key. If v' can establish a (direct or indirect) key with the destination node v, it forwards the request to node v to establish an indirect key.

To show an example, we use the same configuration as in Figure 4.13. When node 1 wants to establish a pairwise key with node 12, it first tries the bridge $\langle 1, 10 \rangle$. If this fails, it tries the bridge $\langle 3, 12 \rangle$. If both bridges fail, it needs to try the bridge $\langle 2, 11 \rangle$. If none of these bridges works, the path key establishment fails. In our later analysis, we will see that it is usually very unlikely that none of those bridges works.

Note that in the above approach, the path key establishment in a cross-group instance has never been used. The reason is that the sensor nodes in a cross group usually spread over the entire deployment field, which may introduce significant communication overhead in path key establishment.

4.3.3 Performance Analysis

For simplicity, we assume that all those in-group and cross-group key pre-distribution instances have the same property functions ($M(D)$, $p_{dk}(D)$, and $p_{cd}(D, x)$). Indeed, this assumption is true for the key pre-distribution techniques in [20, 12, 18, 42] given the same storage overhead, group size, and keying material size. Thus, we use M, p_{dk}, and $p_{cd}(x)$ to represent the three property functions, respectively. Table 4.2 lists the notations that are used frequently in our analysis.

Overhead

Obviously, the storage overhead on a sensor node can be estimated as $2M$. The communication overhead to establish a direct key is the same as the communication overhead to establish a direct key in an in-group or cross-group key pre-distribution instance. When two nodes need to establish an indirect key, there are two cases. If these two nodes are in the same deployment group, the path key establishment only involves the sensor nodes in this deployment group. If these two nodes are in different deployment groups, the path key establishment only involves those in the same deployment group with the source node or the destination node. In other words, the communication is limited in two deployment groups. In addition, we also note that if two sensor nodes in two deployment groups are neighbors, the corresponding deployment groups have high probability of being close to each other, which may reduce the overall communication overhead significantly in their path key establishment.

Table 4.2. Notations

n	number of deployment groups
m	number of nodes in a deployment group
c	number of compromised sensor nodes
M	memory required for one key pre-distribution instance
p_{dk}	probability of having a direct key in a key pre-distribution instance
$p_{cd}(x)$	probability of a direct key being compromised in a key pre-distribution instance when the adversary has randomly compromised x nodes
p_{gdk}	probability of having a direct key in the group-based scheme
$p_{gcd}(x)$	probability of a direct key being compromised in the group-based scheme when the adversary has randomly compromised x nodes
$p_{gci-in}(x)$	probability of an indirect key between two nodes in the same deployment group being compromised when the adversary has randomly compromised x nodes
$p_{gci-cr}(x)$	probability of an indirect key between two nodes in different deployment groups being compromised when the adversary has randomly compromised x nodes

Establishing Direct Keys

Consider a particular sensor node u in the deployment group G_i at position (x', y'). Let A denote its *communication area* in which any other sensor node can directly communication with node u. We assume A is a circle centered at (x', y') with radius R, where R is the radio range of a sensor node. Thus, the average number of sensor nodes in the deployment group G_j that finally reside in A can be estimated as

$$n_{i,j}(x', y') = m \iint_A f(x - x_j, y - y_j) \mathrm{d}x \mathrm{d}y.$$

For any deployment group G_j other than G_i, we know that there is only one sensor node u' in G_j that shares the same cross group G'_k with node u. Thus, the probability of this node u' being deployed in A can be estimated as $\frac{n_{i,j}(x', y')}{m}$. This indicates that among all those sensor nodes deployed in A, the average number of senor nodes that belong to the deployment groups other than G_i but share the same cross group G'_k with node u can be estimated as

$$n'_i(x', y') = \frac{\sum_{j=1, j \neq i}^{n} n_{i,j}(x', y')}{m}.$$

When sensor nodes are evenly distributed in the deployment field, it is possible to further simplify the above equation. Suppose the average number of sensor nodes in the communication range of a sensor node is n_A. We have $\sum_{j=1, j \neq i}^{n} n_{i,j}(x', y') = n_A - n_{i,i}(x', y')$. Thus,

$$n_i'(x', y') = \frac{n_A - n_{i,i}(x', y')}{m}.$$

In addition, the probability of having a direct key between u and any sensor node that shares the same key pre-distribution instance with u is p_{dk}. Thus, the average number of sensor nodes in A that can establish direct keys with node u can be estimated as $(n_{i,i}(x', y') + n_i'(x', y')) \times p_{dk}$. This means that the probability of u having direct keys with its neighbor nodes can be estimated as

$$p_i(x', y') = \frac{(n_{i,i}(x', y') + n_i'(x', y')) \times p_{dk}}{n_A}.$$

Hence, for any node in group G_i, the probability of having direct keys with its neighbor nodes can be estimated as

$$p_{gdk} = \iint_S f(x - x_i, y - x_i) p_i(x, y) \mathrm{d}x \mathrm{d}y,$$

where S denotes the entire deployment field.

p_{gdk} can also be used to estimate the probability of any node in any deployment group having a direct key with its neighbor node when S is an infinite field. For a given deployment field S, we simply configure the deployment point of G_i as its geometric centroid and use the probability of a node in G_i having a direct key with its neighbor node to represent the probability of having a direct key between any two neighbor nodes.

To evaluate the performance of our approach when it is combined with a particular key pre-distribution technique (e.g., the random pairwise keys scheme), we use the following configuration. We assume there are totally 10,000 sensor nodes deployed on a $1000m \times 1000m$ area. These sensor nodes are divided into 100 deployment groups with 100 sensor nodes in each group ($n = m = 100$). We assume sensor nodes are evenly distributed in the deployment field so that the probability of finding a node in each equal size region can be made approximately equal. In other words, the density of sensor nodes is approximately one sensor node per 100 square meters. We always assume the radio range is $R = 40m$. Thus, there are $\frac{\pi \times 40 \times 40}{100} \approx 50.27$ sensor nodes on average in the communication range of a given sensor node. We also set $\sigma = 50m$ in all those deployment distributions $\{f_i(x, y)\}_{i=1,\ldots,n}$.

Figure 4.14 shows the probability of having a direct key between two neighbor nodes under the above configuration. We can see that the probability p_{gcd} increases almost linearly as p_{dk} increases. Since p_{dk} can be made quite large with small storage overhead for a small group of sensor nodes, we expect that the group-based schemes can improve the performance of existing key pre-distribution techniques significantly. To illustrate this point, we investigate the improvements we can achieve by combining the framework with the

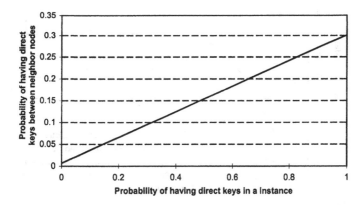

Fig. 4.14. Probability of having a direct key between two neighbor nodes.

basic probabilistic key pre-distribution scheme in [20], the random pairwise keys scheme in [12], and the polynomial-based key pre-distribution in [6]. The result of combination generates three novel key pre-distribution schemes: a *group-based EG* scheme, which combines the framework with the basic probabilistic scheme; a *group-based RK* scheme, which combines the framework with the random pairwise keys scheme; and a *group-based PB* scheme, which combines the framework with the polynomial-based scheme.

For the basic probabilistic key pre-distribution scheme, we assume the key pool size is 100,000. This key pool is divided into 200 small equal size key pools in the group-based EG scheme (500 keys in each small key pool). Each key pre-distribution instance uses a unique key pool. Each sensor node selects the same number of keys from the key pools in its in-group instance and cross-group instance. Figure 4.15 shows that the group-based EG scheme improves the probability of having a direct key between two neighbor sensor nodes significantly when there are severe memory constraints (e.g., 50 keys on each sensor node).

Figure 4.16 compares the probability of having direct keys between neighbor nodes for both the random pairwise keys scheme in [12] and the group-based RK scheme under the same memory constraint. We can clearly see that our framework can significantly improve the probability of having a direct key between two neighbor sensor nodes for the random pairwise keys scheme. This indicates that the group-based RK scheme can support larger sensor networks than the random pairwise keys scheme given the same configuration.

Figure 4.17 shows the probability of having direct keys between neighbor sensor nodes for the group-based PB scheme, the random subset assignment scheme [42], and the grid-based scheme [42]. For all these schemes, we assume the same number of bivariate polynomials in the system and the same number of polynomial shares stored on each sensor node. Specifically, there are 100 de-

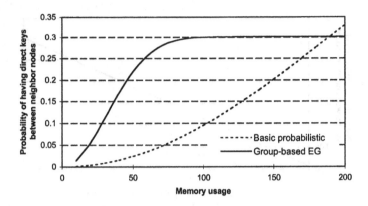

Fig. 4.15. Probability of having a direct key between two neighbor sensor nodes. Memory usage is measured by counting the number of keys stored on each node.

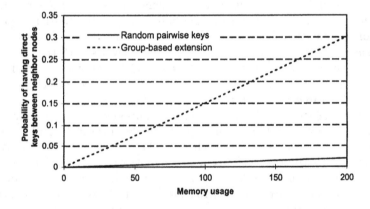

Fig. 4.16. Probability of having a direct key between two neighbor sensor nodes. Memory usage is measured by counting the number of keys stored on each node.

ployment groups and 100 cross groups for the group-based PB scheme. Each of these groups is assigned one unique bivariate polynomial for the corresponding key pre-distribution instance. Each sensor node gets assigned the polynomial shares on its in-group instance and cross-group instance. Similarly, there are 200 bivariate polynomials in the polynomial pools of the random subset assignment scheme and the grid-based scheme. The random subset assignment scheme assigns the polynomial shares of two randomly selected polynomials from the pool to each sensor node, while the grid-based scheme arranges 200 polynomials on a 100×100 grid. We can clearly see that the probability of having a direct key between two neighbor sensor nodes in the group-based

PB scheme is much higher than that in the random subset assignment scheme and the grid-based scheme.

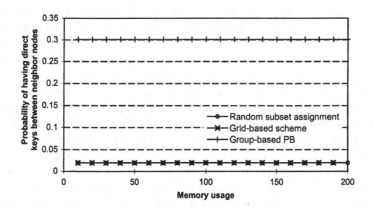

Fig. 4.17. Probability of having a direct key between two neighbor sensor nodes. Memory usage is measured by counting the number of polynomial coefficients stored on each node.

Establishing Indirect Keys

In the following, we estimate the probability of having an indirect key between two neighbor sensor nodes if they cannot establish a direct key.

Obviously, if two neighbor sensor nodes are in the same deployment group G_i, they can follow the path key establishment of D_i to establish an indirect key. We note that a deployment group usually has a limited number of sensor nodes (e.g., 100). Since the nodes in the same deployment group are usually close to each other, a sensor node can easily contact most of the other nodes in the same deployment group. For example, a sensor node can launch a *group flooding*, where only the sensor nodes in the same group participate in the flooding, to contact other nodes. Thus, we believe that it is usually possible to configure the key pre-distribution instance for a deployment group with small storage overhead so that any two sensor nodes in this group can either share a direct key or establish an indirect key at a very high probability with reasonable communication overhead. For example, we employ the random pairwise keys scheme in [12] for a group of 100 sensor nodes and assign 50 keys to each sensor node. In this case, a sensor node can establish a direct key with its neighbor node at a probability of 0.5. After contacting half of the sensor nodes in this group, the probability of finding one node that shares direct keys with both the source and destination nodes can be estimated as

$1 - (1 - 0.5 \times 0.5)^{50} \approx 0.999999$. Hence, we always assume two sensor nodes in the same deployment group can always establish an indirect key.

The situation becomes more complicated if two sensor nodes are in different deployment groups. In this case, they have to find a valid bridge between these two deployment groups to establish an indirect key. Since there are m cross groups, there are m potential bridges. As long as one of them works, the source node can establish an indirect key with the destination node through this bridge. The probability that none of these bridges works can be estimated as $(1 - p_{dk})^m$. Thus, the probability that at least one bridge works, which is equivalent to the probability of having an indirect key between two neighbor nodes in different deployment groups, can be estimated as $1 - (1 - p_{dk})^m$.

Figure 4.18 illustrates the probability of having an indirect key between two neighbor sensor nodes that are in different deployment groups, assuming the same configuration as in Section 4.3.3 for the group-based EG scheme, the group-based RK scheme, and the group-based PB scheme. We can see that two neighbor sensor nodes in different deployment groups can usually establish an indirect key even if there are severe memory constraints on sensor nodes (e.g., 10 keys per sensor node).

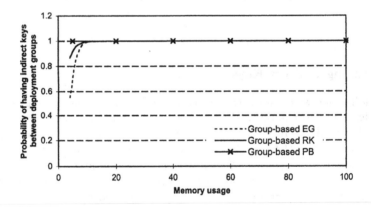

Fig. 4.18. Probability of having indirect keys between sensor nodes in different deployment groups. Memory usage is measured by counting the number of keys or polynomial coefficients stored on each node.

4.3.4 Security Analysis

The main threat we consider in the security analysis is the compromise of sensor nodes. We assume an adversary randomly compromises c sensor nodes in the network. This subsection focuses on the impact of compromised sensor nodes on the direct key establishment and the path key establishment.

Similar to the analysis in the previous subsection, we investigate the security of the proposed framework after combining it with the basic probabilistic key pre-distribution scheme in [20], the random pairwise keys scheme in [12], and the polynomial-based key pre-distribution in [6].

It is easy to see that the grid-based scheme in [42] can be considered as a group-based PB scheme if a row or a column of sensor nodes in the grid are deployed in the same group. This means that the grid-based scheme and the group-based PB scheme have the same security performance against node capture attacks given the same configuration (e.g., storage overhead, network size). Thus, in our later security analysis, we simply skip the security comparison between the grid-based scheme and the group-based PB scheme. On the other hand, we noticed in Figure 4.17 that the group-based PB scheme can achieve much higher probability of establishing direct keys between neighbor sensor nodes than the grid-based scheme. This implies that the group-based PB scheme is more desirable than the grid-based scheme when the group-based deployment model is made possible.

During the evaluation, we always assume that the memory usage at each sensor node is equivalent to storing 100 cryptographic keys. According to the previous configuration, there are $10,000$ sensor nodes in the network, and $n = m = 100$. Thus, for the random pairwise keys scheme, the probability of having a direct key between two neighbor nodes is 0.01, while for the group-based RK scheme, the probability of having a direct key between two neighbor nodes is 0.15 as shown in Figure 4.16.

In addition to the above key pre-distribution schemes, we configure all other schemes in such a way that the probability of having a direct key between two neighbor sensor nodes is 0.3.

- *Basic probabilistic scheme in [20]*: The key pool size is 28,136. Each sensor node randomly selects 100 keys from this pool.
- *Random subset assignment scheme in [42]*: The polynomial pool size is 13, and each polynomial has the degree of 49. Each sensor node randomly selects 2 polynomials from the pool and stores the corresponding polynomial shares.
- *Group-based EG scheme*: The key pool size in each instance is 500. Each sensor node randomly selects 50 keys from its in-group instance and 50 keys from its cross-group instance.
- *Group-based PB scheme*: Each instance includes a 49-degree bivariate polynomial. Each sensor node gets assigned the polynomial shares from its in-group instance and cross-group instance.

Impact on Direct Key Establishment

Consider a direct key between two non-compromised sensor nodes in the same deployment group G_i. Since there are totally c compromised sensor nodes, the probability of j sensor nodes in group G_i being compromised can be estimated as $\frac{c!}{(c-j)!j!}\frac{(n-1)^{c-j}}{n^c}$ for $j \leq m-2$. When j sensor nodes in group G_i are compromised, the probability of this direct key being compromised can be estimated as $p_{cd}(j)$. Hence, the probability of any direct key between two non-compromised sensor nodes in a deployment group being compromised can be estimated as

$$p_{gcd}(c) = \sum_{j=0, j<=c}^{m-2} \frac{c!}{(c-j)!j!}\frac{(n-1)^{c-j}}{n^c}p_{cd}(j)$$

Since $n = m$, the above $p_{gcd}(c)$ can also be used to estimate the probability of a direct key between two non-compromised sensor nodes in the same cross group being compromised.

Figure 4.19 compares the probability of a direct key between two non-compromised sensor nodes being compromised for the basic probabilistic key pre-distribution scheme in [20] and the group-based EG scheme. We can see that the security of direct keys can be significantly improved by applying our framework.

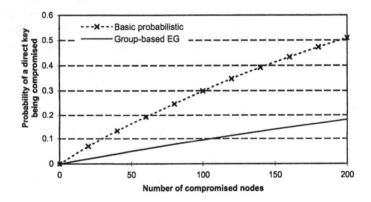

Fig. 4.19. Probability of a direct key between two non-compromised nodes being compromised. Assume the probability of having a direct key between two neighbor nodes is 0.3.

For the random pairwise keys scheme [12], the compromise of sensor nodes does not affect any of the direct keys established between non-compromised sensor nodes ($p_{cd}(j) = 0$) since every key is generated randomly and independently. Thus, if we apply our framework to the random pairwise keys scheme, the resulting scheme still has the perfect security guarantee against node capture attacks ($p_{gcd}(c) = 0$), which means that the compromise of sensor nodes does not affect direct keys between non-compromised nodes. Together with the result in Figure 4.16, we can conclude that our framework can improve the probability of having direct keys between neighbor sensor nodes significantly without sacrificing the security of direct keys.

Figure 4.20 shows the probability of a direct key between two non-compromised sensor nodes being compromised for the group-based PB scheme and the random subset assignment scheme in [42]. We can see that the group-based PB scheme has much better security performance than the random subset assignment scheme in terms of the compromised direct keys.

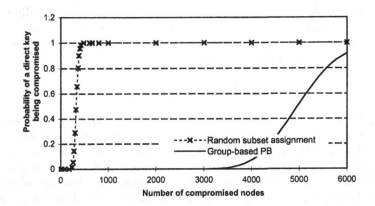

Fig. 4.20. Probability of a direct key between two non-compromised nodes being compromised. Assume the probability of having a direct key between two neighbor nodes is 0.3.

Impact on Path Key Establishment

In the following, we first study the impact of compromised sensor nodes on the indirect keys established between sensor nodes in the same deployment group (in-group indirect keys), and then study the impact of compromised sensor nodes on the indirect keys established between sensor nodes in different deployment groups (cross-group indirect keys).

Note that once the compromised nodes are detected, two non-compromised nodes can always re-establish an indirect key through path key establishment and avoid those compromised sensor nodes or compromised key pre-distribution instances. However, it is usually very difficult to detect compromised sensor nodes. When the compromised nodes cannot be detected, the indirect key between two non-compromised nodes may be disclosed to the attacker without being noticed. In the following analysis, we focus on the probability of a given indirect key between two non-compromised sensor nodes being compromised when the node capture attacks cannot be detected.

Probability of compromised in-group indirect keys: When there are c compromised sensor nodes, the probability of a particular sensor node being compromised can be estimated as $\frac{c}{nm-2}$. According to our earlier analysis, the probability of establishing an in-group indirect key that only involves one intermediate node is usually very high. For simplicity, we assume the in-group indirect key can always be established through one intermediate node. Thus, the establishment of an in-group indirect key involves an intermediate node, a direct key for the link between the source node and the intermediate node, and a direct key for the link between the intermediate node and the destination node. Thus, if the intermediate node and the two direct keys are not compromised, the indirect key is still secure. This means that the probability of an in-group indirect key between two non-compromised nodes being compromised can be estimated as

$$p_{gci-in}(c) = 1 - (1 - \frac{c}{nm-2})(1 - p_{gcd}(c))^2$$

Figure 4.21 shows the probability of an in-group indirect key between two non-compromised nodes being compromised for the group-based EG scheme. It also includes the probability of a given indirect key (involving only one intermediate node) between two non-compromised nodes being compromised for the basic probabilistic scheme in [20]. We can see that the group-based EG scheme has higher security guarantee for the indirect keys between the sensor nodes in the same deployment group.

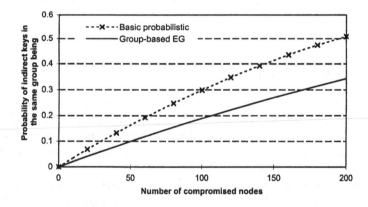

Fig. 4.21. $p_{gci-in}(c)$ for the group-based EG scheme and the probability of an indirect key being compromised for the basic probabilistic scheme. Assume the probability of having a direct key between two neighbor nodes is 0.3.

For the group-based RK scheme, since $p_{gcd}(c) = 0$, we have $p_{gci-in}(c) = \frac{c}{nm-2}$. This means that given the same network size, the probability of an in-group indirect key being compromised for the group-based RK scheme will be equal to the probability of a given indirect key (involving only one intermediate node) being compromised in the random pairwise keys scheme in [12]. However, we note the probability of having a direct key between two neighbor nodes in the random pairwise keys scheme is much lower than that in the group-based RK scheme. In fact, given a large sensor network and small storage overhead, it is very difficult and expensive for the random pairwise keys scheme to establish an indirect key (not to mention the indirect key that involves only one intermediate node) between two neighbor nodes. On the other hand, according to the analysis in Section 4.3.3, we know that the probability of having an indirect key between two neighbor nodes is almost 1 for the group-based RK scheme even if there are severe memory constraints on sensor nodes. Hence, in later discussion, we will also skip the security comparison between these two schemes.

Figure 4.22 shows the probability of an in-group indirect key between two non-compromised nodes being compromised for the group-based PB scheme. It also includes the probability of a given indirect key (involving only one intermediate node) between two non-compromised nodes being compromised for the random subset assignment scheme in [42]. We can see that the group-based PB scheme has much better security performance than the random subset assignment scheme in terms of the compromised indirect keys between nodes in the same deployment group.

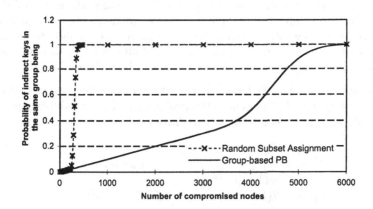

Fig. 4.22. $p_{gci-in}(c)$ for the group-based PB scheme and the probability of an indirect key being compromised for the random subset assignment scheme. Assume the probability of having a direct key between two neighbor nodes is 0.3.

Probability of compromised cross-group indirect keys: Though the establishment of an in-group indirect key involves one intermediate node, the establishment of an indirect key between sensor nodes in different groups may involve up to four intermediate nodes.

Assume the source node u in group G_i wants to setup an indirect key with the destination node v in group G_j. Assume the indirect key is established through a bridge $\langle u', v' \rangle$, where $u' \in G_i$ and $v' \in G_j$. Since the key established between u and v is an indirect key, we have either $u \neq u'$ or $v \neq v'$. Thus, we need to consider the following three cases:

1. *u and v share the same cross group*: The probability of this case can be estimated as $\frac{1}{m}$. In addition, we also note that $u \neq u'$ and $v \neq v'$. Thus, the probability of the path key establishment involving two intermediate nodes can be estimated as p_{dk}^2, which means that u shares a direct key with u', and v shares a direct key with v'. Similarly, the probability of the path key establishment involving three intermediate nodes can be estimated as $2(1 - p_{dk})p_{dk}$, and the probability of the path key establishment involving four intermediate nodes can be estimated as $(1 - p_{dk})^2$.

2. *u and v belong to different cross groups with either $u = u'$ or $v = v'$*: The probability of this case can be estimated as $\frac{m-1}{m}(1 - (1 - p_{dk})^2)$. Similar to the analysis in the first case, the probability of the path key establishment involving one intermediate node can be estimated as p_{dk}, and the probability of the path key establishment involving two intermediate nodes can be estimated as $1 - p_{dk}$.

3. *u and v belong to different cross groups with neither $u' = u$ nor $v' = v$*: The probability of this case can be estimated as $\frac{m-1}{m}(1 - p_{dk})^2$. Similar to the analysis in the first case, the probability of the path key establishment involving two intermediate nodes can be estimated as p_{dk}^2, the probability of the path key establishment involving three intermediate nodes can be estimated as $2(1 - p_{dk})p_{dk}$, and the probability of the path key establishment involving four intermediate nodes can be estimated as $(1 - p_{dk})^2$.

Consider an indirect key established between two sensor nodes in different deployment groups. Let p_i denote the probability of the establishment of this key involving i intermediate nodes; we have

$$
\begin{cases}
p_1 = \frac{m-1}{m}[1 - (1 - p_{dk})^2]p_{dk} \\
p_2 = \frac{1}{m}p_{dk}^2 + \frac{m-1}{m}[(1 - (1 - p_{dk})^2)(1 - p_{dk}) \\
\quad + (1 - p_{dk})^2 p_{dk}^2] \\
p_3 = 2(1 - p_{dk})p_{dk}[\frac{1}{m} + \frac{m-1}{m}(1 - p_{dk})^2] \\
p_4 = \frac{1}{m}(1 - p_{dk})^2 + \frac{m-1}{m}(1 - p_{dk})^2(1 - p_{dk})^2
\end{cases}
$$

When the path key establishment involves i intermediate nodes, the indirect key will be still secure if all of these i nodes and the related $i + 1$ direct keys are not compromised. Thus, for an indirect key that involves i intermediate nodes, the probability of it being compromised can be estimated as $1 - (1 - p_{gcd}(c))^{i+1}(1 - \frac{c}{nm-2})^i$. Hence, the probability of a cross-group indirect key between two non-compromised sensor nodes being compromised can be estimated as

$$p_{gci-cr}(c) = \sum_{i=1}^{4} p_i \times [1 - (1 - p_{gcd}(c))^{i+1}(1 - \frac{c}{nm - 2})^i].$$

Figure 4.23 shows the probability of a cross-group indirect key between two non-compromised sensor nodes being compromised for the group-based EG scheme. It also includes the probability of an indirect key (involving only one intermediate node) between two non-compromised nodes being compromised for the basic probabilistic scheme [20]. We can see that the security of these two schemes are very close to each other in terms of the indirect keys between sensor nodes in different deployment groups.

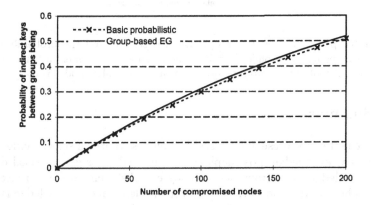

Fig. 4.23. $p_{gci-cr}(c)$ for the group-based EG scheme and the probability of an indirect key being compromised for the basic probabilistic scheme. Assume the probability of having a direct key between two neighbor nodes is 0.3.

Figure 4.24 shows the probability of a cross-group indirect key between two non-compromised sensor nodes being compromised for the group-based PB scheme. It also includes the probability of an indirect key (involving only one intermediate node) between two non-compromised nodes being compromised for the random subset assignment scheme in [42]. We can still see that the group-based PB scheme has much better security performance than the random subset assignment scheme in terms of the indirect keys between nodes in different deployment groups.

According to the above security analysis and the performance analysis in the previous subsection, we can easily conclude that the proposed framework can significantly improve the security as well as the performance of existing key pre-distribution techniques.

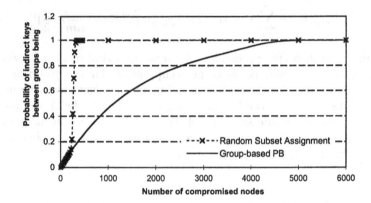

Fig. 4.24. $p_{gci-cr}(c)$ for the group-based PB scheme and the probability of an indirect key being compromised for the random subset assignment scheme. Assume the probability of having a direct key between two neighbor nodes is 0.3.

4.4 Summary

In this chapter, we presented several techniques to utilize sensor nodes' prior deployment knowledge, post deployment knowledge, or group-based deployment knowledge to improve pairwise key establishment in static sensor networks. The analysis shows that when certain deployment knowledge is available, we can improve the performance of existing key pre-distribution techniques significantly.

Several research directions are worth further study, including detailed performance evaluation through simulation and the implementation of these techniques on real sensor platforms.

5

Secure Localization

In this chapter, we first briefly introduce the background knowledge for secure localization in sensor networks and then develop two techniques to tolerate malicious attacks against the location discovery in wireless sensor networks. Our first technique, an *attack-resistant Minimum Mean Square Estimation (MMSE)* technique, is based on the observation that malicious location references introduced by attacks are intended to mislead a sensor node about its location and thus are usually inconsistent with the benign ones. To take advantage of this observation, our attack-resistant MMSE method identifies malicious location references by examining the inconsistency among location references (which is indicated by the mean square error of location estimation) and defeats malicious attacks by removing such malicious data. Our second technique, a *voting-based location estimation* method, quantizes the deployment field into a grid of cells and has each location reference "vote" on the cells in which the node may reside. This method then filters out the effects of malicious location references by choosing the geometric centroid of the cell(s) with the highest vote as the estimated location. We have implemented the proposed techniques on MICA2 motes [14] running TinyOS [26] and evaluated the security and performance through both simulation and field experiments. The experimental results indicate that the proposed techniques are not only effective but also practical.

We also introduce a suite of techniques to detect and remove compromised beacon nodes that supply misleading location information to the regular sensors, aiming at providing secure location discovery services in wireless sensor networks. We develop an efficient method to detect malicious beacon signals using redundant beacon nodes in the sensing field. The basic idea is to take advantage of the (known) locations of beacon nodes and the constraints that these locations and the measurements (e.g., distance, angle) derived from their beacon signals must satisfy to detect malicious beacon signals. With this method, we propose a serial of techniques to detect replayed beacon signals to avoid false positives in detecting malicious beacon nodes. We also present a simple method to reason about the suspiciousness of each beacon node and revoke malicious beacon nodes based on the distributed detection results from beacon nodes. The analysis and simulation results show that the proposed techniques are practical and effective in detecting and removing malicious beacon nodes.

5.1 Localization in Sensor Networks

Existing localization schemes either employ *range-based* methods , which use the exact measurements obtained in stage one, or *range-free* methods , which only need the existences of beacon signals in stage one. Typical techniques to obtain the measurements between two nodes include Received Signal Strength Indicator (RSSI) , Time of Arrive (ToA) , Time Difference of Arrive (TDoA) , and Angle of Arrive (AoA) .

The basic idea of localization in sensor networks can be illustrated through a simple example. As shown in Figure 5.1, non-beacon node O wants to estimate its location. In the first phase, Node O measures the distances d_1, d_2, and d_3 to beacon node A, B and C respectively, based on the received beacon signals. In the second phase, node O lists the following equations according to the measured distances and the locations of beacon nodes.

$$\begin{cases} f_1 = d_1 - \sqrt{(x-x_1)^2 + (y-y_1)^2} \\ f_2 = d_2 - \sqrt{(x-x_2)^2 + (y-y_2)^2} \\ f_3 = d_3 - \sqrt{(x-x_3)^2 + (y-y_3)^2} \end{cases}$$

A typical way to solve the above equations is to use MMSE (Minimum Mean Square Estimation) so that $F = f_1^2 + f_2^2 + f_3^2$ is minimized. In general, when non-beacon nodes receive beacon signals from more than three beacon nodes, the accuracy of location estimation can be improved by solving a serial of such equations using MMSE.

$$\begin{cases} f_1 = d_1 - \sqrt{(x-x_1)^2 + (y-y_1)^2} \\ f_2 = d_2 - \sqrt{(x-x_2)^2 + (y-y_2)^2} \\ \vdots \qquad\qquad \vdots \\ f_m = d_m - \sqrt{(x-x_m)^2 + (y-y_m)^2} \end{cases}$$

With these equations, we can use the technique in [71] to estimate the location of sensor nodes.

The above example shows the basic idea of a range-based localization scheme. Range-based localization schemes in sensor networks include those in [71, 72, 56, 52, 16]. Savvides et al. developed AHLoS localization protocol based on Time Difference of Arrive [71]. Extension of this work can be found in [72]. Doherty et al. presented a localization scheme based on connectivity-induced constraints and the relative angle between neighbors [16]. Angle of Arrive is also used to develop localization schemes in [56] and [52].

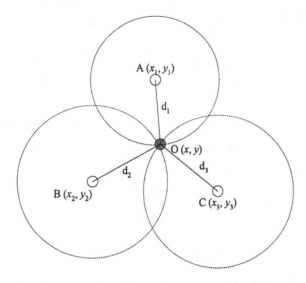

Fig. 5.1. An example of localization method. Nodes A,B and C are beacon nodes, and node O is a non-beacon node.

Range-free based schemes are proposed to provide location estimation services for those applications with less required location precision [9, 57, 51, 25]. Bulusu, Heidemann and Estrin proposed a simple range-free, coarse grained localization scheme where each sensor estimated its location by centering the locations contained in the received signals [9]. Niculescu and Nath proposed using the minimum hop count and the average hop size to estimate the distance between two nodes and then use multilateration to estimate the location [57].

5.2 Pitfalls of Current Localization Schemes under Attacks

All of the current localization schemes become vulnerable when there are malicious attacks. In all these schemes, the accuracy of location estimation depends on the accuracy of the origins of the beacon signals (which are assumed to be the locations in the beacon packets) and certain measurements obtained from the beacon signals, including distances and/or angles in range-based schemes, and the existence of beacon signals in range-free schemes. Though the above measurements are directly obtained from the physical signals, the locations of the beacon signals' origins can be easily forged, as discussed earlier (See Figure 5.2). As a result, a malicious attacker may introduce large errors when a node estimates its location.

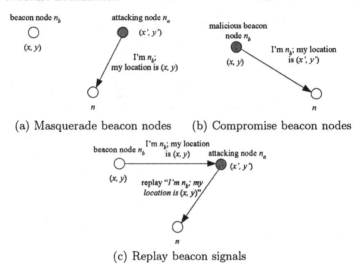

(a) Masquerade beacon nodes   (b) Compromise beacon nodes

(c) Replay beacon signals

Fig. 5.2. Attacks against location discovery schemes

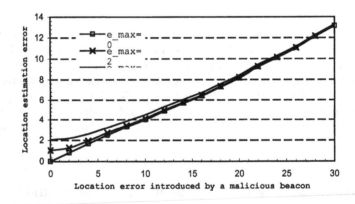

Fig. 5.3. Location estimation error of a MMSE-based method in presence of malicious attacks

Most of the localization schemes for sensor networks have certain ability to tolerate measurement errors (e.g., by averaging the effect of problematic location references over all location references). For example, Minimum Mean Square Estimation (MMSE) has been used in most of the range-based and some range-free localization schemes [71, 72, 56, 52, 16, 57] to improve the accuracy of location estimation when a sensor node has redundant location references. However, these methods cannot properly handle malicious location references, which typically include very large errors not seen in natural faults.

To demonstrate the impact of malicious attacks, we performed an experiment through simulation with the MMSE-based location estimation method in [71]. We used 9 honest beacon nodes and 1 malicious beacon node randomly deployed in a $30m \times 30m$ field. The node that estimates location is positioned at the center of the field. The malicious beacon node always declares a false location x meters away from its real location, where x is a parameter representing the location error. To model the distance measurement error, we assume such an error is uniformly distributed between $-e_{max}$ and e_{max}. Figure 5.3 shows the location estimation error (i.e., the distance between a sensor's real location and the estimated location) introduced by the malicious beacon node. We can clearly see that the malicious node affects the estimated location significantly by declaring incorrect locations. Indeed, the location estimation error at the sensor node increases almost linearly with the error introduced by the malicious node. Since an attacker can introduce arbitrarily large errors by declaring false locations in beacon packets, the above result implies that the attacker can introduce arbitrarily large errors into a non-beacon node's location estimation.

Such malicious attacks will generate similar impacts on the other localization schemes. This is because an attacker may introduce arbitrary errors into the location estimation process, while all the existing localization techniques assume bounded errors, which are only true in benign environments. As discussed in the Introduction, such attacks cannot be simply prevented by cryptographic techniques due to the threat of compromised nodes and replay attacks.

5.3 Attack-Resistant Location Estimation

In this section, we present two approaches to dealing with malicious attacks against location discovery in wireless sensor networks. The first approach is extended from the minimum mean square estimation (MMSE). It uses the mean square error as an indication to identify and remove malicious location references. The second one adopts an iteratively refined voting scheme to survive malicious location references introduced by attackers.

In the following, we clarify several assumptions of the proposed techniques before describing these methods.

5.3.1 Assumptions

We assume that all the beacon packets are authenticated. With authentication, beacon packets forged by external attackers who do not have access to keying materials can be easily filtered out without being considered for location estimation. Authentication is practical in the current sensor networks. For example TinySec [33], developed by the UC Berkeley group, can provide packet authentication on MICA motes [14] running TinyOS [26].

We also assume that all beacon nodes are uniquely identified. In other words, each non-beacon node can identify the sender of each beacon packet, based on the cryptographic key used to authenticate the packet. There are two possible ways to do this. First, each non-beacon node may share a pairwise key with each beacon node from which it may receive a beacon signal. Several pairwise key pre-distribution schemes [20, 12, 18, 42] have been proposed to establish pairwise keys in sensor networks; they can be potentially used for our application. Second, the beacon nodes may use a broadcast authentication protocol such as μTESLA [63]. In either case, a non-beacon node that receives a beacon signal can identify the generator of the message authentication code (MAC) in the beacon packet and then the sender.

Finally, we assume that each non-beacon node uses at most one location reference derived from the beacon signals sent by each beacon node. As a result, even if a beacon node is compromised, the attacker that has access to the compromised key can only introduce at most one malicious location reference.

5.3.2 Attack-Resistant Minimum Mean Square Estimation (MMSE)

Intuitively, a location reference introduced by a malicious attack is aimed at misleading a sensor node about its location. Thus, it is usually "different" from benign location references. When there are redundant location references, there must be some "inconsistency" between the malicious location references and the benign ones. To take advantage of this observation, we propose to use the "inconsistency" among the location references to identify the malicious ones, and discard them before finally estimating the locations at sensor nodes.

We assume that a sensor node uses a MMSE-based method to estimate its own location. Thus, most of the current range-based localization techniques can be used with this technique. To specifically harness the above observation about inconsistent location references, we first estimate the sensor node's location with the MMSE-based method, and then assess whether the estimated location could be derived from a set of consistent location references. If yes, we accept the estimated location; otherwise, we identify and remove the most "inconsistent" location reference, and repeat the above process. This process may continue until we find a set of consistent location references or it is not possible to find such a set.

For simplicity, we assume the distances measured from beacon signals are used for location estimation. (This approach can certainly be modified to accommodate other measurements such as angles.) For the sake of presentation, we denote a location reference obtained from a beacon signal as a triple $\langle x, y, \delta \rangle$, where (x, y) is the location of the beacon node declared in the beacon packet, and δ is the distance measured from the beacon signal.

We use the mean square error ς^2 of the distance measurements based on the estimated location as an indication of the degree of inconsistency, since all the

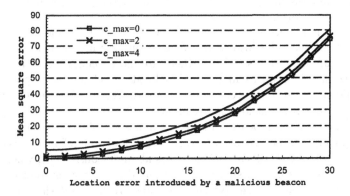

Fig. 5.4. The effect of malicious attacks on the mean square error ς^2

MMSE-based methods estimate a sensor node's location by (approximately) minimizing this mean square error.

Definition 5.1. *Given a set of location references* $\mathcal{L} = \{\langle x_1, y_1,\ \delta_1 \rangle, \langle x_2, y_2,\ \delta_2 \rangle, ..., \langle x_m, y_m, \delta_m \rangle\}$ *and a location* $(\tilde{x}_0, \tilde{y}_0)$ *estimated based on* $\mathcal{L}$, *the mean square error of this location estimation is*

$$\varsigma^2 = \sum_{i=1}^{m} \frac{(\delta_i - \sqrt{(\tilde{x}_0 - x_i)^2 + (\tilde{y}_0 - y_i)^2})^2}{m}.$$

Intuitively, the more inconsistent a set of location references is, the greater the corresponding mean square error should be. To gain further understanding, we plot in Figure 5.4 the mean square error ς^2 corresponding to the location estimation errors shown in Figure 5.3. As Figure 5.4 shows, if a malicious beacon node increases the location estimation error by introducing greater errors in beacon packets, it also increases the mean square error ς^2 at the same time. This further demonstrates that the mean square error ς^2 is potentially a good indicator of inconsistent location references.

In this chapter, we choose a simple, threshold-based method to determine whether a set of location references is consistent. Specifically, a set of location references $\mathcal{L} = \{\langle x_1, y_1, \delta_1 \rangle, \langle x_2, y_2, \delta_2 \rangle, ..., \langle x_m, y_m, \delta_m \rangle\}$ obtained at a sensor node is τ-*consistent* w.r.t. a MMSE-based method if the method gives an estimated location $(\tilde{x}_0, \tilde{y}_0)$ such that the mean square error of this location estimation

$$\varsigma^2 = \sum_{i=1}^{m} \frac{(\delta_i - \sqrt{(\tilde{x}_0 - x_i)^2 + (\tilde{y}_0 - y_i)^2})^2}{m} \leq \tau^2.$$

The threshold τ is clearly a critical parameter in our approach. We will discuss how to determine τ in Section 5.3.2. In the following, we first describe the algorithm for the attack-resistant MMSE method, assuming the threshold τ is already set properly.

Though it is necessary to remove malicious location references to defeat malicious attacks, the MMSE-based location estimation methods can deal with measurement errors better if there are more benign location references. Thus, we should keep as many benign location references as possible once the malicious location references are removed. This implies we should get the largest set of consistent location references.

Given a set $\mathcal{L}$ of n location references and a threshold τ, a naive approach to computing the largest set of τ-consistent location references is to check all subsets of $\mathcal{L}$ with i location references about τ-consistency, where i starts from n and drops until a subset of $\mathcal{L}$ is found to be τ-consistent or it is not possible to find such a set (when i becomes 3). Thus, if the largest set of consistent location references consists of m elements, a sensor node has to use the MMSE method $\binom{n}{m}/2 + \binom{n}{m+1} + \cdots + \binom{n}{n}$ times on average to find the right one. If $n = 10$ and $m = 5$, a sensor node needs to perform the MMSE method for about $\binom{10}{5}/2 + \binom{10}{6} + \binom{10}{7} + \binom{10}{8} + \binom{10}{9} + \binom{10}{10} = 512$ times. It is certainly desirable to reduce the computation for resource-constrained sensor nodes.

To reduce the computation on sensor nodes, we adopt a greedy algorithm , which is simple but suboptimal. This greedy algorithm works in rounds. It starts with the set of all location references in the first round. In each round, it first verifies if the current set of location references is τ-consistent. If yes, the algorithm outputs the estimated location and stops. Optionally, it may also output the set of location references. Otherwise, it considers all subsets of location references with one fewer location reference and chooses the subset with the least mean square error as the input to the next round. This algorithm continues until it finds a set of τ-consistent location references or it is not possible to find such a set. The details of this greedy algorithm are given below.

1. The algorithm starts with a set $\mathcal{L}$ of n ($n > 3$) location references. The algorithm uses a MMSE-based location estimation method to estimate the location using $\mathcal{L}$ and compute the mean square error ς^2. If $\varsigma^2 < \tau^2$, the algorithm outputs the estimated location and returns SUCCESS. Otherwise, it goes to the next step.
2. Assume there are i location references left in set $\mathcal{L}$. If $i = 3$, the algorithm returns FAIL and stops. Otherwise, for each ($i - 1$)-element subset of $\mathcal{L}$, it estimates a location with the MMSE-based method and computes the corresponding mean square error. It then chooses the subset with the least mean square error. If this value is less than τ^2, the algorithm outputs the corresponding location estimate and returns SUCCESS. Otherwise, it replaces $\mathcal{L}$ with the chosen subset and repeats this step.

The greedy algorithm significantly reduces the computational overhead in sensor nodes. To continue the earlier example, a sensor node only needs to perform MMSE operations for about $1 + 10 + 9 + 8 + 7 + 6 + 5 + 4 = 50$ times using this algorithm, which is about 10% of the cost introduced by the naive approach mentioned earlier. In general, a sensor node needs to use a MMSE-based method for at most $1 + n + (n-1) + \cdots + 4 = 1 + \frac{(n-3)(n+4)}{2}$ times.

Determining Threshold τ

As discussed earlier, the threshold τ is a critical parameter in our method. In the following, we investigate how to determine the value of τ.

Our basic idea is to study the distribution of the mean square error ς^2 when there are no malicious attacks, and use this information to help determine the threshold τ. Before we investigate the distribution of ς^2, we first clarify what are *benign* location references in normal situations.

Definition 5.2. *A location reference $\langle x, y, \delta \rangle$ at a sensor node is considered as* benign *if it satisfies $|\delta - \sqrt{(x - x_0)^2 + (y - y_0)^2}| \leq \epsilon$, where (x_0, y_0) is the real location of the sensor node and ϵ is the maximum measurement error (which occurs when measuring the distance δ from the physical beacon signal).*

Intuitively, Definition 5.2 says that a location reference is benign if the distance between the node's real location and the location claimed by the corresponding beacon signal is not very different from the distance measured from the beacon signal. Intuitively, a benign location reference represents a location reference obtained from a beacon signal when there are no malicious attacks; the error introduced by a benign location reference is mainly due to the measurement of the physical signals, which is bounded by ϵ. A malicious attacker can certainly introduce benign location references by having small errors in beacon packets; however, this will not generate big impact on the location estimation as suggested by Figure 5.3. Thus, such location references can be subsumed into the benign ones.

All the localization techniques are aimed at estimating a location as close to the sensor node's real location as possible. Thus, we may assume that the estimated location is very close to the real location when there are no attacks. Next, we derive the distribution of the mean square error ς^2 using the real location as the estimated location and compare it with the distribution obtained through simulation when there are location estimation errors.

When there are no malicious attacks, the location estimation error is all introduced by the measurements of beacon signals. For a location reference $L_i = \langle x_i, y_i, \delta_i \rangle$, the measurement error e_i can be computed as $e_i = \delta_i - \sqrt{(x_0 - x_i)^2 + (y_0 - y_i)^2}$, where (x_0, y_0) is the real location of the sensor node. Assuming that the measurement errors introduced by different benign location references are independent and identical, we can get the distribution of the mean square error through the following Lemma.

Lemma 5.3. *Let* $\{e_1, ..., e_m\}$ *be a set of identical, independent random variables and* μ, σ *be the mean and variance of the random variable* e_i^2, $1 \leq i \leq m$. *If the estimated location of a sensor node is its real location, the probability distribution of* ς^2 *is*

$$\lim_{m \to \infty} F[\varsigma^2 \leq \varsigma_0^2] = \Phi(\frac{m\varsigma_0^2 - m\mu}{\sigma\sqrt{m}}),$$

where $\Phi(x)$ *is the probability that a standard normal random variable is less than* x.

Proof. For a sensor node, we model the measurement error of location reference i as a probability density function $f_i(e_i)$, where $e_i \leq \epsilon$. Thus, we have the following equation:

$$\varsigma^2 = \sum_{i=1}^{m} \frac{(\delta_i - \sqrt{(x_0 - x_i)^2 + (y_0 - y_i)^2})^2}{m} = \sum_{i=1}^{m} \frac{e_i^2}{m}$$

The cumulative distribution function can thus be calculated by

$$F(\varsigma^2 \leq \varsigma_0^2) = F(\sum_{i=1}^{m} e_i^2 \leq m\varsigma_0^2) \tag{5.1}$$

Since $\{e_1^2, e_2^2, \cdots, e_m^2\}$ are independent random variables with mean μ and variance σ^2, according to the central limit theorem, we have

$$\lim_{m \to \infty} P(\frac{S_m - m\mu}{\sigma\sqrt{m}} \leq x) = \Phi(x),$$

where $S_m = \sum_{i=0}^{m} (e_i^2)$. Together with Equation 5.1, we have

$$\begin{aligned}
\lim_{m \to \infty} F(\varsigma^2 \leq \varsigma_0^2) &= \lim_{m \to \infty} F(S_m \leq m\varsigma_0^2) \\
&= \lim_{m \to \infty} P(\frac{S_m - m\mu}{\sigma\sqrt{m}} \leq \frac{m\varsigma_0^2 - m\mu}{\sigma\sqrt{m}}) \\
&= \Phi(\frac{m\varsigma_0^2 - m\mu}{\sigma\sqrt{m}})
\end{aligned}$$

Let us further assume a simple model for measurement errors, where the measurement error is evenly distributed between $-\epsilon$ and ϵ. Then the mean and the variance for e_i are 0 and $\frac{\epsilon^2}{3}$, respectively, and the mean and the variance for any e_i^2 are $\frac{\epsilon^2}{3}$ and $\frac{4\epsilon^4}{45}$, respectively. Let $\varsigma_0 = c \times \epsilon$; we have $F(\varsigma^2 \leq (c \times \epsilon)^2) = \Phi(\frac{\sqrt{5m}(3c^2-1)}{2})$.

Lemma 5.3 describes the probability distribution of ς^2 based on a sensor's real location. Though it is different from the probability distribution of ς^2 based on a sensor's estimated location, it can be used to approximate such distribution in most cases.

Figure 5.5 shows the probability distribution of ς^2 derived from Lemma 5.3 and the simulated results using sensors' estimated locations. We can see

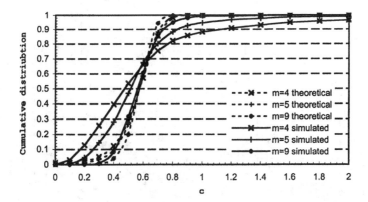

Fig. 5.5. Cumulative distribution function for the mean square error of location estimation on benign location references. Let $c = \frac{\varsigma_0}{\epsilon}$.

that when the number of location references (m) is large (e.g., $m = 9$), the theoretical result derived from Lemma 5.3 is very close to the simulation results. However, when m is small (e.g., $m = 4$), there are observable differences between the theoretical results and the simulation.

The reason is twofold. First, our theoretical analysis is based on the central limited theorem, which is only an approximation of the distribution when m is a large number; it may not catch the real distribution when m is small (e.g. 4). Second, we used the MMSE-based method proposed in [71] in the simulation, which estimates a node's location by only *approximately* minimizing the mean square error. (Otherwise, the value of ς^2 for benign location references should never exceed ϵ^2.)

Figure 5.5 gives three hints about the choice of the threshold τ. First, when there are enough number of benign location references, a threshold less than the maximum measurement error is enough. For example, when $m = 9$, $\tau = 0.8\epsilon$ can guarantee the nine benign location references are considered consistent with high probability. Besides, a large threshold may lead to the failure to filter out malicious location references. Second, when m is small (e.g. 4), the cumulative probability becomes flatter and flatter when $c > 0.8$. This means that setting a large threshold τ for small m may not help much to guarantee the consistency test for benign location references; instead, it may give an attacker high chance to survive the detection. Third, the threshold cannot be too small; otherwise, a set of benign location references has high probability to be determined as a non-consistent reference set.

Based on the above observations, we propose to choose the value for τ with a hybrid method. Specifically, when the number of location references is large (e.g., more than 8), we determine the value of τ based on the theoretical results

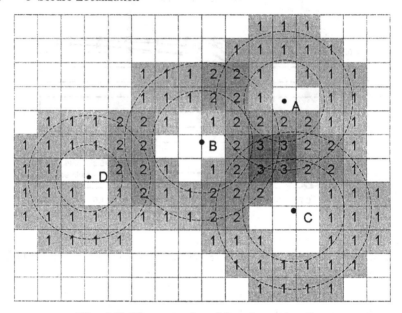

Fig. 5.6. The voting-based location estimation

derived from Lemma 5.3. Specifically, we choose a value of τ corresponding to a high cumulative probability (e.g., 0.9). When the number of location references is small, we perform simulation to derive the actual distribution of the mean square error and then determine the value of τ accordingly. Since there are only a small number of simulations to run, we believe this approach is practical.

We have evaluated the attack-resistant MMSE method through both simulation and field experiments, which will be reported in Sections 5.3.5 and 5.3.6, respectively.

5.3.3 Voting-Based Location Estimation

In this approach, we have each location reference "vote" on the locations at which the node of concern may reside. To facilitate the voting process, we quantize the target field into a grid of cells and have each sensor node determine how likely the node is located in each cell based on each location reference. We then select the cell(s) with the highest vote and use the "center" of the cell(s) as the estimated location.

After collecting a set of location references, a sensor node should determine the target field. The node does so by first identifying the minimum rectangle that covers all the locations declared in the location references and then extending this rectangle by R_b, where R_b is the maximum transmission range of a beacon signal. This extended rectangle forms the target field, which contains all possible locations for the sensor node. The sensor node then divides this rectangle into M small squares (cells) with the same side length L, as illustrated in Figure 5.6. (The node may need to further extend the target field to have square cells.) The node then keeps a voting state variable for each cell, initially set to 0.

Consider a benign location reference $\langle x, y, \delta \rangle$. The node that has this location reference must be in a ring centered at (x, y), with the inner radius $\max\{\delta - \epsilon, 0\}$ and the outer radius $\delta + \epsilon$. For the sake of presentation, we refer to such a ring a *candidate ring (centered) at location* (x, y) . For example, in Figure 5.6, the grey ring centered at point A is a candidate ring at A, which is derived from the location reference with the declared beacon node location at A.

For each location reference $\langle x, y, \delta \rangle$, the sensor node identifies the cells that overlap with the corresponding candidate ring and increments the voting variables for these cells by 1. After the node processes all the location references, it chooses the cell(s) with the highest vote and uses its (their) geometric centroid as the estimated location of the sensor node.

Overlap of Candidate Rings and Cells

A critical problem in the voting-based approach is how to determine whether a candidate ring overlaps with a cell. We discuss how to determine this efficiently below.

Assume we need to determine whether the candidate ring at A overlaps with the cell shown in Figure 5.7(a). We denote the minimum distance from a point in the cell to point A as $d_{min}(A)$, and the maximum distance from a point in the cell to point A as $d_{max}(A)$. It is easy to see that the candidate ring does not overlap with the cell only when $d_{min}(A) > r_o$ or $d_{max}(A) < r_i$, where $r_i = \max\{0, \delta - \epsilon\}$ and $r_o = \delta + \epsilon$ are the inner and the outer radius of the candidate ring, respectively. Thus, if we can compute $d_{min}(A)$ and $d_{max}(A)$ for a cell and a candidate ring centered at A, we can easily determine whether they overlap.

To compute d_{min} and d_{max}, we divide the target field into 9 regions based on the cell, as shown in Figure 5.7(b). It is easy to see that given the center of any candidate ring, we can determine the region in which it falls with at most 6 comparisons between the coordinates of the center and those of the corners of the cell. When the center of a candidate ring is in region 1 (e.g., point A in Figure 5.7(b)), it can be shown that the closest point in the cell to A is the upper left corner, and the farthest point in the cell from A is the lower right corner. Thus, $d_{min}(A)$ and $d_{max}(A)$ can be calculated accordingly. These two distances can be computed similarly when the center of a candidate ring falls into regions 3, 7, and 9.

Consider point B that falls into region 2. Assume the coordinate of point B is (x_B, y_B). We can easily see that $d_{min}(B) = y_B - y_2$. The computation of $d_{max}(B)$ is a little more complex. We first need to check if $x_B - x_1 > x_2 - x_B$. If yes, the farthest point in the cell from B must

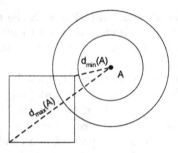

(a) Overlap of a ring and a cell

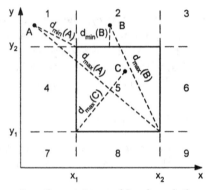

(b) Computing the minimum (d_{min}) and the maximum distance (d_{max}) between a cell and the center of a ring

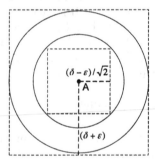

(c) Limiting the examinations of cells

Fig. 5.7. Determine whether a ring overlaps with a cell

be the lower left corner of the cell. Otherwise, the farthest point in the cell from B should be the lower right corner of the cell. Thus, we can get $d_{max}(B) = \sqrt{(\max\{x_B - x_1, x_2 - x_B\})^2 + (y_B - y_1)^2}$. These two distances can be computed similarly when the center of a candidate ring falls into regions 4, 6, and 8.

Consider a point C that falls into region 5. Obviously, $d_{min}(C) = 0$ since point C itself is in the cell. Assume the coordinate of point C is (x_C, y_C). The farthest point in the cell from C must be one of its corners. Similarly to the above case for point B, we may check which point is farther away from C by checking $x_C - x_1 > x_2 - x_C$ and $y_C - y_1 > y_2 - y_C$. As a result, we get $d_{max}(C) = \sqrt{(\max\{x_C - x_1, x_2 - x_C\})^2 + (\max\{t_C - y_1, y_2 - y_C\})^2}$.

According to the above discussion, we can determine whether a cell and a candidate ring overlap with at most 10 comparisons and a few arithmetic operations. To prove the correctness of the above approach only involves elementary geometry; we omit the proofs from the chapter.

For a given candidate ring, a sensor node does not have to check all the cells for which it maintains voting state variables. As shown in Figure 5.7(c), with simple computation, the node can get the outer bounding box centered at A with side length $2(\delta + \epsilon)$. The node only needs to consider the cells that intersect with or fall inside the bounding box. Moreover, the node can get the inside bounding box with simple computation, which is centered at A with side length $\sqrt{2}(\delta - \epsilon)$, and all the cells that fall into this bounding box need not be checked.

Iterative Refinement

The number of cells M (or equivalently, the quantization step L) is a critical parameter for the voting-based algorithm. It has several implications to the performance of our approach. First, the larger M is, the more state variables a sensor node has to keep, and thus the more storage is required. Second, the value of M (or L) determines the precision of location estimation when there are no attacks. The larger M is, the smaller each cell will be. As a result, a sensor node can determine its location more precisely based on the overlap of the cells and the candidate rings. Finally, the value of M also has direct implication for the ability to tolerate malicious attacks. To demonstrate this, Figure 5.8 shows the cumulative probability of location estimation errors when M has different values, where there is a malicious location reference declaring a wrong location with a 20m error. It is shown that a larger M does result in higher probability to achieve a given precision under malicious attacks.

Due to the resource constraints on sensor nodes, the granularity of the partition is usually limited by the memory available for the voting state variables on the nodes. This puts a hard limit on the accuracy of location estimation. To address this problem, we propose an *iterative refinement* of the above *basic algorithm* to achieve fine accuracy with reduced storage overhead.

In this iterative refinement version, the number of cells M (or equivalently, the side length of each cell L) is chosen according to the memory constraint in a sensor node, so that the node has enough memory to have a voting state

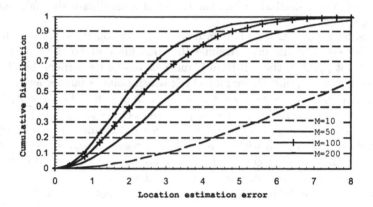

Fig. 5.8. Cumulative probability distribution of location estimation error for different partition configurations. Assume the location error created by a malicious beacon node is $20m$, and the measurement error is $\epsilon = 4m$.

variable (normally one byte) for each cell. After the first round of the algorithm, the node may find one or more cells having the largest vote. To improve the accuracy of location estimation, the sensor node then identifies the smallest rectangle that contains all the cells having the largest vote, and performs the voting process again. For example, in Figure 5.6, the same algorithm will be performed in a rectangle which exactly includes the 4 cells having 3 votes. Note that in a later iteration of the basic voting-based algorithm, a location reference does not have to be used if the corresponding candidate ring does not overlap with any of the cells with the highest vote.

In a later iteration, due to a smaller rectangle to quantize, the size of each cell can be reduced. This leads to a higher precision and better ability to deal with malicious attacks. Moreover, a malicious location reference will most likely be discarded, since the corresponding candidate ring usually does not overlap with those derived from benign location references. For example, in Figure 5.6, the candidate ring centered at point D will not be used in the second iteration since it does not overlap with the 3-vote cells.

The iterative refinement process should terminate when a desired precision is reached or the estimation cannot be refined. The former condition can be tested by checking the side length of a cell in an iteration. In other words, if in one iteration the side length of each cell is less than a predefined threshold (which indicates the desired precision of location estimation), the algorithm should output the geometric centroid of the cell(s) with the highest vote as the estimated location of the node. However, there may be cases that the iterative refinement cannot further refine the precision of the location estimation. This can be determined also by checking the side length of each cell; if this side length remains the same as the last iteration, the location estimation cannot be further refined. The algorithm should simply stop and output the estimated location obtained in the last iteration as the result. It is easy to see that the algorithm will fall into either of these two cases, and thus will always terminate. In practice, we may set the desired precision to 0 in order to get the best precision.

The proposed iterative voting-based location estimation allows a resource-constrained sensor node to gradually refine its location estimation to a desired precision. When the majority of the location references are benign, it is highly likely that the few malicious location references are filtered out in the second iteration since these malicious location references are intended to mislead the sensor node about its location and very possibly do not overlap with the cells with the highest vote. Even if the majority of the location references are malicious, they have to be consistent with each other in order not to be discarded. This greatly increases the level of coordination and thus the difficulty of malicious attacks.

5.3.4 Security Analysis

Based on the descriptions of the proposed techniques, it is easy to see that when there are more benign location references than the malicious ones introduced by attackers, the effect of the malicious ones will be removed from the final location estimation. Moreover, the attacker must carefully control the beacon packets and the beacon signals in order to keep them consistent with each other and to overwhelm the benign beacon signals. Indeed, to defeat the attack-resistant MMSE approach, the attacker needs to control the declared locations in beacon packets and the physical features (e.g., signal strength) of beacon signals so that the malicious location references are considered consistent and overwhelm the benign location references. To defeat the voting-based approach, the attacker needs to control beacon packets and beacon signals similarly so that the cell containing the attacker's choice gets more votes than the cell(s) containing the sensor node's real location.

To further understand the difficulty of malicious attacks, let us consider specific approaches an attacker may use to launch the above attacks. An attacker has two ways to transmit the above malicious beacon signals. First, the attacker may compromise beacon nodes and use the compromised beacon nodes or the compromised keys to generate malicious beacon signals. Since all beacon packets are authenticated, and a sensor node uses at most one location reference derived from the beacon signals sent by each beacon node, the attacker needs to compromise more beacon nodes than the benign beacon nodes from which a target sensor node may receive beacon signals, besides carefully crafting the forged beacon signals.

Second, the attacker may launch wormhole [29] or replay attacks to tunnel benign beacon signals from one area to another. In this case, the attacker does not have to compromise any beacon nodes. However, the attacker will have to face several other difficulties. Let us first consider wormhole attacks. First, the attacker has to use multiple wormholes so that the beacon signals from the wormholes suppress the benign ones a sensor node receives locally. Moreover, the attacker has to carefully arrange the starting points and the ending points of the wormholes to maintain the consistency among the tunneled beacon signals. Finally, the wormhole attacks have to bypass potential wormhole detections such as those proposed in [29, 28]. When replay attacks are used, the attacker will have to face the same difficulties in coordinating multiple beacon signals.

Based on the above analysis, we can see that though it is theoretically possible for an attacker to defeat the proposed location estimation techniques, it takes substantial effort to launch such attacks in practice. Thus, we can conclude that the proposed techniques significantly increase the security of location estimation in wireless sensor networks.

In the following sections, we further examine the security, performance, and practicality of the proposed techniques through simulation and field experiments.

5.3.5 Simulation Evaluation

This section presents the simulation results of both attack-resistant schemes proposed in the previous section. The evaluation focuses on the performance under different choices of parameters and the improvement on the accuracy of location estimation in hostile environments. A comparison between these two schemes is also provided.

Three attack scenarios are considered in the evaluation. The first scenario considers a single malicious location reference that declares a wrong random location that is e meters away from the beacon node's real location. In the second scenario, there are multiple malicious location references, and each of them declares a wrong random location that is e meters away from the beacon node's real location. In the third scenario, multiple colluding malicious location references are considered. In this case, the malicious location references declare false locations by coordinating with each other to create a virtual location e meters away from the sensor's real location. Thus, a set of malicious location references may appear to be consistent to a victim node.

In all simulations, a set of benign beacon nodes and a few malicious beacon nodes are randomly deployed in a $30m \times 30m$ target field. The non-beacon sensor node is located at the center of this target field. We assume the maximum transmission range of beacon signals is $R_b = 22m$, so that the non-beacon node can receive the beacon signal from every beacon node located in the target field. We assume the entire deployment field is much larger than this target field so that an attacker can create a very large location estimation error inside the deployment field. Each malicious beacon node declares a false location according to the three attack scenarios discussed above. We assume a simple distance measurement error model. That is, the distance measurement

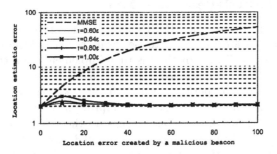

(a) a single malicious location references

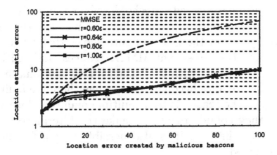

(b) 3 random malicious location references

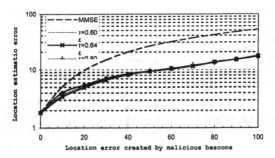

(c) 3 colluding malicious location references

Fig. 5.9. Performance of attack-resistant MMSE (with 9 benign location references)

error is uniformly distributed between $-\epsilon$ and ϵ, where the maximum distance measurement error ϵ is set to $\epsilon = 4m$.

Evaluation of Attack-Resistant MMSE

In our simulation, we use the MMSE-based method proposed in [71], which we call the *basic MMSE method*, to perform the basic location estimation. Our attack-resistant MMSE method is then implemented on the basis of the basic MMSE method, as discussed in Section 5.3.2.

We first evaluate the performance of the attack-resistant MMSE scheme under different threshold τ and show the improvement over the basic MMSE method in the presence of malicious attacks. In the simulation, we selected 4 thresholds according to Figure 5.5: 0.6ϵ, 0.64ϵ, 0.8ϵ, and ϵ. These four choices guarantee 9 benign location references are considered consistent with probabilities of 0.6, 0.8, 0.999, and 1, respectively.

Figure 5.9 illustrates the performance of the attack-resistant MMSE method under the four selected thresholds when there are malicious location references. For comparison, Figure 5.9 also includes the performance of the basic MMSE-based method. This figure shows that when there are malicious attacks, the attack-resistant MMSE reduces the location estimation error significantly compared with the basic MMSE-based method. It is worth noting that the performance becomes worse when there are multiple malicious location references. The reason is that multiple malicious location references, especially when they collude together, make the filtering of malicious location references more difficult. It is also possible that a few benign location references are removed.

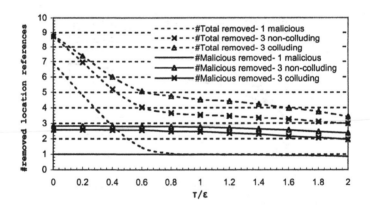

Fig. 5.10. Average number of removed location references v.s. the value of threshold τ. Assume there are 9 benign location references and each malicious location reference introduces 40m location error.

Figure 5.10 shows the average number of location references removed by the attack-resistant MMSE method under different thresholds when malicious location references introduce 40m location errors. As indicated by this figure, malicious location references are not always removed, and benign location references may be mistakenly removed – especially when there are multiple colluding malicious location references. However, as the threshold is around ϵ, fewer benign location references are removed. When the threshold increases beyond ϵ, not only are fewer benign location references removed, but also some malicious ones are kept. This is consistent with our earlier discussion. In practice, a trade-off between removing malicious and benign location references needs to be made.

Despite the fact that some benign location references may be removed and some malicious ones may be used for location estimation, the attack-resistant MMSE method still performs much better than the basic MMSE method, as shown in Figure 5.9.

Evaluation of Voting-Based Scheme

The partition granularity M and the desired precision of location estimation S are two important parameters in the voting-based scheme. Both have impacts on the accuracy of the estimated location. However, S is a parameter that allows a user to get less precision when a high precision is not necessary. Thus, in the simulation, we set $S = 0$ to get the minimum location estimation error achievable by this method.

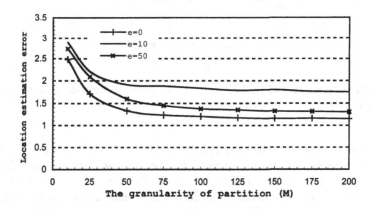

Fig. 5.11. Performance for different M (e: error introduced by a malicious location reference)

We first study the impact of parameter M on the voting-based method. Figure 5.11 shows the performance of the voting-based scheme with different values of M when there is only one malicious location reference. We can see that the location estimation error initially decreases when M increases but does not decrease much when M is greater than 100. Moreover, the parameter M also has implications in computational cost. Since the voting-based method is finally reduced to checking whether a candidate ring derived from a location reference overlaps with the cells in the grid, we use the number of cells being examined as an indication of the computational cost. Figure 5.12 shows the computational costs of the voting-based method for different values of M when there is one malicious location reference. As this figure shows, the computational cost increases almost linearly with the value of M. When there are no or more malicious location references, the computational cost will increase similarly as M increases. Based on these results, we set $M = 100$ in the later simulations to trade-off the accuracy with the storage and computation overhead.

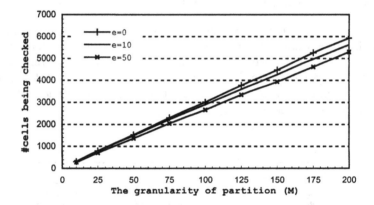

Fig. 5.12. Computational cost for different M (e: error introduced by a malicious location reference)

Now let us examine the performance of the voting-based scheme against malicious attacks. Figure 5.13 compares the accuracy of the basic MMSE method and our voting-based scheme under different types of attacks. We can clearly see that the accuracy of location estimation is improved significantly in our scheme. In addition, unlike the attack-resistant MMSE scheme, the voting-based scheme can tolerate multiple (colluding or non-colluding) malicious location references more effectively.

Note that the curves for the voting-based scheme in Figure 5.13 have a bump when the location error introduced by malicious location references is around 10m. This is because the malicious location references are not significantly different from the benign location references around this point, and our scheme cannot completely shield the effect of malicious location refer-

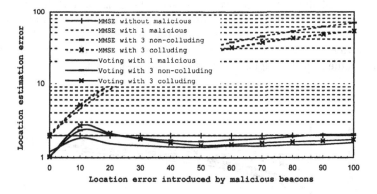

Fig. 5.13. Performance of the voting-based scheme ($M = 100$ and $S = 0$)

ences. Nevertheless, the attacker will not be able to introduce large location estimation errors if they do not introduce large location errors. As a result, the location estimation errors are always bounded, even if there are malicious attacks.

Comparison between Two Proposed Schemes

Now let us compare the attack-resistant MMSE and the voting-based methods. Based on the earlier results, we choose threshold $\tau = 0.8\epsilon$ for the attack-resistant MMSE, and set $M = 100$ and $S = 0$ for the voting-based scheme.

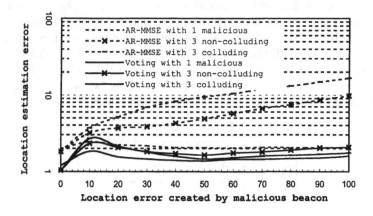

Fig. 5.14. Attack-resistant MMSE scheme v.s. the voting-based scheme

Figure 5.14 shows that the voting-based scheme is more resilient than the attack-resistant MMSE scheme. Moreover, Figure 5.15 shows that when the distance measurement errors are small, the attack-resistant MMSE scheme could be more accurate than the voting-based scheme. However, the voting-based scheme still has higher performance in most cases.

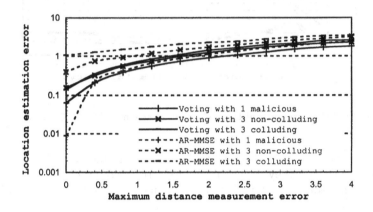

Fig. 5.15. Performance under different distance measurement errors (assume that the location error introduced by each malicious location references is 10m)

5.3.6 Implementation and Field Experiments

We have implemented both proposed schemes on TinyOS [26] (version 1.1.0), an operating system for networked sensors. These implementations are targeted at MICA2 motes [14] running TinyOS. As discussed in Section 5.3.5, the attack-resistant MMSE implementation is based on the MMSE method proposed in [71]. However, our implementation of the basic MMSE method is simplified by only using the location coordinates (without the ultrasound propagation speed, which is not necessary in our study).

Scheme	ROM (bytes)	RAM (bytes)
MMSE	2034	286
AR-MMSE	3226	396
Voting-Based	4488	174

Table 5.1. Code size for different schemes (assume a maximum of 12 location references; $M = 100$)

Table 5.1 gives the code size (ROM and RAM) for these implementations on MICA2 platform. Table 5.1 is obtained by assuming at most 12 location references. More location references will increase the RAM size of the program, but the increased RAM is only required to save the additional location references. These numbers indicate the proposed schemes are practical on the current generation of sensor nodes, such as MICA2 motes.

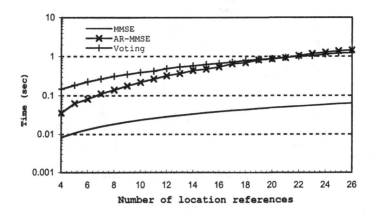

Fig. 5.16. Average execution time on MICA2 motes ($\epsilon = 4$m, $\tau = 0.8\epsilon$, $M = 100$ and $S = 0$)

Figure 5.16 shows the average execution time of the basic MMSE, the attack-resistant MMSE, and the voting-based schemes on MICA2 motes. These data are collected by counting the number of CPU clock cycles spent on location estimation. The location references used in the experiment are generated from the simulation in Section 5.3.5. We can see that the basic MMSE method requires the least execution time. The attack-resistant MMSE scheme has less computational cost than the voting-based scheme when the number of location references is small; however, when there are large numbers of location references (e.g., 20), it takes the voting-based method less time to finish than the attack-resistant MMSE method.

To further study the feasibility of the proposed techniques, we performed an outdoor field experiment. In the field experiment, eight MICA2 motes were deployed in a 10×10 target field, where each unit of distance is 4 feet, as shown in Figure 5.17. The sensor node with ID 0 is configured as a non-beacon sensor node which is located at the center of the field. All the other sensor nodes are configured as beacon nodes.

We considered three attack scenarios in the field experiment. In the first scenario, the beacon node with ID 1 is configured as a malicious beacon node. This node always declares a location e feet away from from its real location,

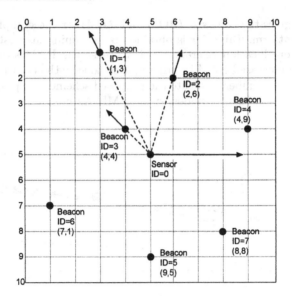

Fig. 5.17. Target area of field experiment.

in the direction away from the non-beacon node. (The arrow marked on the beacon node 1 in Figure 5.17 shows this direction.) In the second scenario, beacon nodes 1, 2 and 3 are configured as malicious beacon nodes. Similar to the first scenario, each of these three nodes declares a location that is e feet away from its real location in the directions away from the non-beacon node. In the third scenario, three malicious beacon nodes 1, 2, and 3 work together to create a virtual location. Each of these three nodes declares a false location by increasing its horizontal coordinate by e feet. This actually creates a virtual location in the horizontal axis that is e feet away from the non-beacon node's real location. This is illustrated in Figure 5.17 by the horizontal arrow starting from the non-beacon node.

To measure the distance between sensor nodes, we use a simple RSSI (Received Signal Strength Indicator) based technique. Note that the Active Message protocol in the current version of TinyOS provides a reading in the *strength* field for the MICA2 platform. This value is returned in every received packet and can be used to compute the received signal strength. To take advantage of this field, we performed an experiment before the actual field experiment in the test field to estimate the relationship between the values of this field and the distance between two nodes. For each given distance, we computed the average of these values on 20 observations. By doing this, we built a table which contains distances (from 0 to 30 feet with step size of 2 feet) and the corresponding average readings. During the field experiments, when a sensor node receives 20 packets from a beacon node, it computes the average of the values contained in the strength field and estimates the distance with interpolation according to this table. For example, if the average reading v falls in between two adjacent points (v_i, d_i) and (v_{i+1}, d_{i+1}) in the table, the sensor computes the corresponding distance $d = d_i + \frac{(v-v_i) \times (d_{i+1}-d_i)}{v_{i+1}-v_i}$.

In the field experiment, the observed maximum distance measurement error is about 4 feet. Thus, we set ϵ to 4 feet.

Figure 5.18 shows the performance of the attack-resistant methods as well as the basic MMSE method in the field experiment. For the first two attack scenarios, we can see that the attack-resistant methods can filter out or tolerate malicious location references quite effectively. The performance in the third scenario is worse than in the first two cases. The reason is that the non-beacon node has only 4 benign location references, but 3 colluding location references. This is almost the worst situation we can deal with by purely using the location references. However, we can still see that the location estimation error drops when the location errors introduced by the malicious attacks are above certain thresholds. Overall, the location estimation errors caused by malicious attacks are bounded when the proposed techniques are used, while the errors can be arbitrarily large when the basic MMSE method is used.

The field experiment further confirms that the proposed attack-resistant location estimation methods are efficient and effective. Moreover, it shows these techniques are practical on the current generation of sensor networks.

5.4 A Detector for Malicious Beacon Nodes

In hostile environments, a compromised beacon node or an attacking node that has access to compromised cryptographic keys may send out malicious beacon signals that include incorrect locations or manipulate the beacon signals so that a receiving node obtains, for example, incorrect distance measurements. Sensor nodes that use such beacon signals for location determination may estimate incorrect locations. In this section, we first describe a simple but effective method to detect malicious beacon signals. With this method, we then develop techniques to filter out replayed beacon signals and thus detect malicious beacon nodes.

We assume that two communicating nodes share a unique pairwise key. A number of random key pre-distribution schemes (e.g., [12, 42, 18]) can be used for this purpose. We assume that a beacon node cannot tell if it is communicating with a beacon or non-beacon node simply from the radio signal or the key used to authenticate the packet. We also assume that communication is two way; that is, if node A can reach node B, then node B can reach node A as well. Moreover, we assume beacon signals are unicasted to non-beacon nodes, and every beacon packet is authenticated (and potentially encrypted) with the pairwise key shared between two communicating nodes. Hence, beacon packets forged by external attackers that do not have the right keys can be easily filtered out.

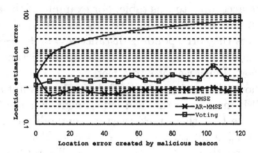

(a) A single malicious location reference

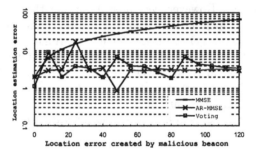

(b) 3 non-colluding malicious location references

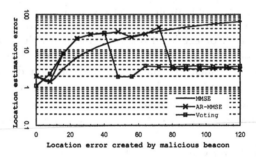

(c) 3 colluding malicious location references

Fig. 5.18. Performance of the attack-resistant schemes in the field experiment ($M = 100$ and $S = 0$ for the voting-based scheme; $\tau = 0.8\epsilon = 3.2$ feet for attack-resistant MMSE)

We assume location estimation is based on the distances measured from beacon signals (through, e.g., RSSI). Nevertheless, our approach can be easily revised to deal with location estimation based on other measurements.

5.4.1 Detecting Malicious Beacon Signals

The technique to detect malicious beacon signals is the basis of detecting malicious beacon nodes. The basic idea is to take advantage of the (known) locations of beacon nodes and the constraints that these locations and the measurements (e.g., distance, angle) derived from the beacon signals must satisfy to detect malicious beacon signals.

A beacon node can perform detection on the beacon signals it hears from other beacon nodes. For the sake of presentation, we call the node making this detection the *detecting (beacon) node* , and the node being detected the *target (beacon) node* . Note that if a malicious beacon node knows that a detecting beacon node is requesting for its beacon signal, it can send out a normal beacon signal that does not lead to incorrect location estimation and thus pass the detection mechanism without being noticed. To deal with this problem, the detecting node uses a different node ID, called *detecting ID* , during the detection. This ID should be recognized as a non-beacon node ID. The detecting node also has all keying materials related to this ID so that it can communicate securely with other beacon nodes as a non-beacon node. To increase the probability of detecting a malicious beacon node, we may allocate multiple detecting IDs as well as the related keying materials to each beacon node. With the help of these detecting IDs, it is very difficult for an attacker to distinguish the requests generated by detecting beacon nodes from those generated by non-beacon nodes when sensor nodes are densely deployed. If sensor nodes have certain mobility and/or the detecting node can carefully craft its request message (e.g., adjust the transmission power in RSSI technique), it will become even more difficult for the attacker to determine the source of a request message. For simplicity, we assume that the attacker cannot tell if a request message is from a beacon node or a non-beacon node.

The proposed method works as follows. The detecting node n first sends a request message to the target node n_a as a non-beacon node. Once the target node n_a receives this message, it sends back a beacon packet (beacon signal) that includes its own location (x', y'). The detecting node n then estimates the distance between them from the beacon signal upon receiving it. Since the detecting node n knows its own location, it can also calculate the distance between them based on its own location (x, y) and the target node's location (x', y'). The detecting node n then compares the estimated distance and the calculated one. If the difference between them is larger than the maximum distance error, the detecting node can infer that the received beacon signal must be malicious. Figure 5.19 illustrates this idea.

A potential problem in the above method is that even if the calculated distance is consistent with the estimated distance, it is still possible that the beacon signal comes from a compromised beacon node or is replayed by an attacking node. However, a further investigation reveals that this will not generate impact on location estimation. Consider a malicious beacon node that declares a location (x', y'). If the estimated distance from its beacon

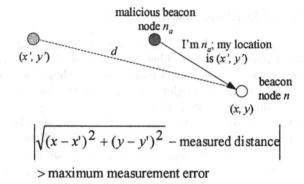

$$\left| \sqrt{(x - x')^2 + (y - y')^2} - \text{measured distance} \right|$$

$$> \text{maximum measurement error}$$

Fig. 5.19. Detect malicious beacon signals

signal is consistent with the calculated one, it is equivalent to the situation where a benign beacon node located at (x', y') sends a benign beacon signal to the requesting node. In fact, to mislead the location estimation at a non-beacon node, the attacker has to manipulate its beacon signal and/or beacon packet to make the estimated distance inconsistent with the calculated one. This manipulation will certainly be detected if the requesting node happens to be a detecting node.

5.4.2 Filtering Replayed Beacon Signals

Suppose a beacon signal from a target node is detected to be malicious, it is still not clear if this node is malicious since an attacker may replay a previously captured beacon signal. However, if we can determine that a malicious beacon signal indeed comes directly from this target node, this target node must be malicious. Thus, it is necessary to filter out as many replayed beacon signals between benign beacon nodes as possible in the detection.

A beacon signal may be replayed through a *wormhole attack* [29], where an attacker tunnels packets received in one part of the network over a low latency link and replays them in a different part [29]. Wormhole attacks generate big impacts on the security of many protocols (e.g., localization, routing). A number of techniques have been proposed recently to detect such attacks, including geographical leashes [29], temporal leashes [29], and directional antenna [28]. These techniques can be used to filter out beacon signals replayed through a wormhole.

A beacon signal received from a neighbor beacon node may also be replayed by an attacking node. We call such replayed beacon signals *locally replayed beacon signals* . Most wormhole detectors cannot deal with such attacks, since they can only tell if two nodes are neighbor nodes. It is possible to use temporal leashes [29] to filter out locally replayed beacon signal since replaying a beacon signal may introduce delay that is detectable with temporal leashes. However, this technique requires a secure and tight time synchronization and large memory space to store authentication keys. Instead, we study the effectiveness of using round trip time to filter out locally replayed beacon signals and demonstrate that using round trip time does not require the time synchronization method but can detect locally replayed beacon signals effectively.

Replayed Beacon Signals from Wormholes

We assume that there is a wormhole detector installed on every beacon and non-beacon node. This wormhole detector can tell whether two communicating nodes are neighbor nodes or not with certain accuracy. The purpose of the following method is to filter out the replayed beacon signals due to the wormhole between two benign beacon nodes that are far away from each other. An observation regarding such replayed beacon signals is that the distance between the location of the detecting node and the location contained in the beacon packet is larger than the communication range of the target node. Thus, we combine the wormhole detector with the location information in the following algorithm.

Once a beacon signal is detected to be a malicious beacon signal, the detecting node begins to verify if it is replayed through a wormhole with the help of the wormhole detector. The detecting node first calculates the distance to the target beacon node based on its own location and the location declared in the beacon packet. If the calculated distance is larger than the radio communication range of the target node and the wormhole detector determines that there is a wormhole attack, the beacon signal is considered as a replayed beacon signal and is ignored by the detecting node. Otherwise, the beacon signal will go through the process to filter locally replayed signals in Section 5.4.2.

Let us briefly study the effectiveness of this method. Since a malicious target node can always manipulate its beacon signals to convince the detecting node that there is a wormhole attack and they are far from each other even if they are neighbor nodes, it is possible that the beacon signal from a malicious target node is removed. Fortunately, non-beacon nodes in the network are also equipped with this wormhole detector. This means that a malicious target node cannot convince all detecting nodes that there are wormhole attacks and at the same time convince all non-beacon nodes that there are no wormhole attacks so that its beacon signals are not removed by non-beacon nodes. This is because a malicious beacon node does not know if a requesting node is a detecting beacon node.

It is also possible that a replayed beacon signal through a wormhole from a benign target node is not removed. The reason is that the wormhole detector cannot guarantee that it can always detect wormhole attacks.

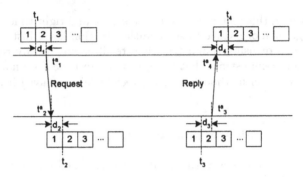

Fig. 5.20. Round trip time

Locally Replayed Beacon Signals

The method to filter out locally replayed beacon signals is based on the observation that the replay of a beacon signal introduces extra delay. In most cases, this delay is large enough to detect whether there is a locally replayed beacon signal through the round trip time (RTT) between two neighbor nodes. In the following, we first investigate the characteristics of RTT between two neighbor sensor nodes in a typical sensor network, and then use this result to filter out locally replayed signals between benign beacon nodes.

To remove the uncertainty introduced by the MAC layer protocol and the processing delay , we measure the RTT in the following way. As shown in Figure 5.20, the sender sends a request message to the receiver, and the receiver responds with a reply message. t_1 is the time of finishing sending the first byte of the request from a sender, t_2 is the time of finishing receiving the first byte of this request at a receiver, t_3 is the time of finishing sending the first byte of the reply from the receiver, and t_4 is the time of finishing receiving the first byte of this reply at the original sender. The sender estimates RTT by computing $RTT = (t_4 - t_1) - (t_3 - t_2)$, where t_4 and t_1 are available at the sender, and $t_3 - t_2$ can be obtained from the receiver by exchanging messages.

Characteristics of RTT between neighbor nodes: We may perform experiments on an actual sensor platform to obtain the characteristics of RTT. To gain further insights and examine our approach, we performed experiments on MICA2 motes [14] running TinyOS [26]. For simplicity, we assume the same type of sensor nodes in the sensor network. Nevertheless, our technique can be easily extended to deal with different types of nodes in the network.

In the experiment, t_1 is measured by recording the time right after the communication module (CC1000) moves the second byte of the request message to the SPDR register , which is used to store the byte being transmitted over the radio channel. In other words, t_1 is the time of finishing shifting the first byte of the request message out of this register. Assume the absolute time of finishing sending the first byte of the request message is t_1^a. We have $t_1 + d_1 = t_1^a$, where d_1 is the delay between shifting the data byte out of the

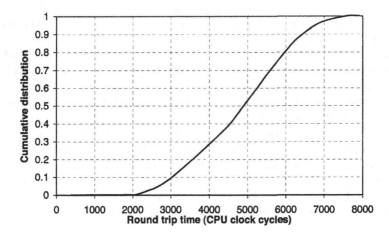

Fig. 5.21. Cumulative distribution of round trip time

SPDR register and finishing sending this byte over the radio channel. Similarly, we have $t_3 + d_3 = t_3^a$. Similarly, t_2 is measured by recording the time right after the first byte of the request message is ready at the SPDR register. Assume the absolute time of finishing receiving this byte from the radio channel is t_2^a. We have $t_2 = t_2^a + d_2$, where d_2 is the delay between receiving this byte from the radio channel and reading this byte from the SPDR register. Similarly, we have $t_4 = t_4^a + d_4$. Since the radio signal travels at the speed of light, we have $t_4^a - t_1^a - (t_3^a - t_2^a) = \frac{2D}{c}$, where D is the distance between two neighbor nodes and c is the speed of light. Thus, we have $RTT = d_1 + d_2 + d_3 + d_4 + \frac{2D}{c}$.

Note that d_1, d_2, d_3 and d_4 are mainly affected by the underlying radio communication hardware. Since two neighbor nodes are usually close to each other, the value of $\frac{2D}{c}$ in the above equation is negligible. Hence, the RTT measured by computing $RTT = (t_4 - t_1) - (t_3 - t_2)$ is not affected by the MAC protocol or any processing delay. This means that the distribution of RTT should be within a narrow range. Let F denote the cumulative distribution function of RTT when there are no replay attacks, x_{min} denote the maximum value of x such that $F(x) = 0$, and x_{max} denote the minimum value of x such that $F(x) = 1$.

Figure 5.21 shows the cumulative distribution of RTT when there are no replay attacks. We use one CPU clock cycle as the basic unit to measure the time. This figure is derived by measuring RTT 100,000 times. The result shows that $x_{min} = 1,951$ and $x_{max} = 7,506$. Since the transmission time of one bit is about 384 clock cycles, we can detect any replayed signal if the delay introduced by this replay is longer than the transmission time of $\frac{7506 - 1951}{384} \approx 14.5$ bits.

The detector for locally replayed beacon signals: With RTT's cumulative distribution, we can detect locally replayed signals between benign beacon nodes effectively. The basic idea is to check if there is any significant difference between the observed RTT and the range of RTT derived during our experiments. For example, if the observed RTT at the requesting node is larger than the maximum RTT in Figure 5.21, it is very likely that the reply signal is replayed. The following local replay detector will be installed on every beacon and non-beacon node.

The requesting node u communicates with a beacon node v following the request-reply protocol shown in Figure 5.20. As a result, node u can compute $RTT = (t_4 - t_1) - (t_3 - t_2)$. There are two cases: (1) When $RTT \leq x_{max}$, the beacon signal is considered as not locally replayed. If the requesting node is a detecting node, it will report an alert when the beacon signal is detected to be malicious. If the requesting node is a non-beacon node, this beacon signal will be used in its location estimation. (2) When $RTT > x_{max}$, this beacon signal is considered as locally replayed and will be ignored by the requesting node.

When the target node is a benign beacon node and is a neighbor of the detecting node, but the beacon signal is replayed by a malicious node, the detecting node will report an alert if the delay introduced by the locally replayed signal is less than the transmission time of 14.5 bits data. However, this is very difficult for the attacker to achieve since the attacker has to replay the beacon signal to the detecting node when the target node is still sending its beacon signal. This implies that the attacker has to physically shield every signal to the detecting node and replay the intercepted packet at the same time. When the target benign beacon node is not a neighbor node of the detecting node, the detecting node will report an alert if the delay introduced by the undetected wormhole attack is less than the transmission time of 14.5 bits data. Note that this implies this replayed beacon signal has bypassed the wormhole detector.

Note that the purpose of the above method is to filter the replayed beacon signals between benign beacon nodes to avoid false positives. This method becomes trivial when the target node is a malicious beacon node, since it can easily convince a detecting node that the beacon signal is locally replayed and thus prevent the detecting node from reporting an alert. However, the malicious target node cannot convince all detecting nodes that the beacon signals are locally replayed and at the same time convince all non-beacon nodes that its beacon signals are not locally replayed so that its beacon signals are accepted by non-beacon nodes.

Security and Performance Analysis

Theoretically, the proposed techniques can be used to provide security for any existing localization scheme based on location references from beacon nodes. However, when TDoA technique is used for measuring distances to beacon nodes, the proposed techniques do not work as effectively as in other techniques (e.g., RSSI, ToA, and AoA) since it is usually more difficult to protect ultrasound signals – especially when ultrasound signals cannot carry data packets.

In some cases, a non-beacon node may become a beacon node to supply location references once it discovers its own location. Localization errors may accumulate when more and more non-beacon nodes turn into beacon nodes. However, there are still constraints between estimated measurements and calculated measurements; otherwise, it is impossible to estimate locations with required accuracy. With these constraints, we can still apply the proposed detector to catch possible malicious beacon nodes, though the specific solutions need further investigation.

Overheads: Since beacon signals are unicasted from beacon nodes to their neighbor non-beacon nodes, our techniques sacrifice a certain amount of communication overhead for security. This trade-off is practical, since location estimation only needs to be done once for each non-beacon node in most cases, and a sensor node (beacon or non-beacon node) usually only needs to communicate with a few other nodes within its communication range. The computation and storage overheads are mainly introduced by key establishment protocol and cryptographic operations.

False positives: Our techniques cannot prevent a malicious detecting node from reporting alerts against other beacon nodes. The techniques are aimed at reducing the probability of a benign beacon node reporting alerts against other benign beacon nodes and increasing the probability of a benign beacon node reporting alerts against malicious beacon nodes.

For simplicity, we assume that when a node A is sending a beacon signal to its neighbor node B during time period T, node B either receives the original signal or receives nothing (in case of collision) at the end of T. Thus, the delay of replaying a signal between two neighbor nodes is at least the transmission time of one entire packet, which is typically much larger than 14.5 bits. This means that our detector can always detect locally replayed beacon signals between two benign neighbor nodes. Hence, the situation where a benign beacon node reports an alert against another benign beacon node only happens when (1) they are not neighbor nodes, (2) the attacker creates a wormhole between them, (3) this wormhole cannot be detected by the detecting node, and (4) the delay introduced by this wormhole is less than the transmission time of 14.5 bits. Assume the detection rate of the wormhole detector is p_d. The probability that a replayed beacon signal through a wormhole from a benign beacon node is not removed can be estimated by $1 - p_d$. Thus, the probability of a benign beacon node reporting an alert on another benign beacon node is at most $1 - p_d$ if there is a wormhole between them, and 0 otherwise.

Detection rate (P_r): The detection rate, which is the probability of a malicious target node being detected by a detecting node, is an important metric to evaluate the performance of our detector. Assume a malicious beacon node u sends normal beacon signals to a fraction p_n of the requesting nodes, convinces a fraction p_w of requesting nodes that its beacon signals are replayed from wormholes and convinces a fraction p_l of requesting nodes that its beacon

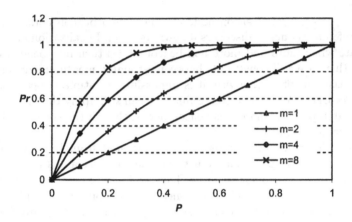

Fig. 5.22. Relationship between P_r and P.

signals are locally replayed. We also assume the malicious beacon node u behaves in the same way for the same requesting node, which is the best strategy for the node to avoid being detected. Thus, using one detecting ID, a benign detecting node v that hears beacon signals from this malicious node u will detect malicious beacon signals with a probability of $1 - p_n$. If a malicious beacon signal from malicious node u is detected, the probability of going through the process of filtering locally replayed beacon signals is $1 - p_w$. During the process of filtering locally replayed beacon signals, the probability of node v reporting an alert against a malicious node u is $1 - p_l$. Hence, the probability of malicious node u being detected by node v can be estimated by $(1 - p_n)(1 - p_w)(1 - p_l)$. When each detecting node has m detecting IDs. The probability P_r of a malicious beacon node being detected by a benign detecting node can be estimated by $P_r = 1 - (1 - (1 - p_n)(1 - p_w)(1 - p_l))^m$.

We denote P as the probability that (1) a requesting non-beacon node receives a malicious beacon signal from a malicious beacon node, and (2) this malicious beacon signal is not removed by the replay detector. For a requesting non-beacon node w, the probability of hearing malicious beacon signals from node u is $(1 - p_n)$. If w receives a malicious beacon signal, the probability of going through the detection of locally replayed signals is $1 - p_w$. During the detection of locally replayed signals, the probability of this malicious beacon signal not being filtered out is $1 - p_l$. Since the above three events are independent from each other, the probability P can be estimated by $P = (1 - p_n)(1 - p_w)(1 - p_l)$. Thus, we have $P_r = 1 - (1 - P)^m$.

Figure 5.22 shows the relationship between the detection rate P_r and P. It indicates that an attacker cannot increase P without increasing the probability of being detected. On the other hand, a benign detecting node can always increase m to have higher detection rate P_r.

5.4.3 Revoking Malicious Beacon Nodes

With the detector in the previous section, a detecting beacon node may report alerts about other suspicious beacon nodes. In this section, we propose to use the base station to further remove malicious beacon nodes from the network to reduce their impact on the location discovery service. We assume that the base station has mechanisms to revoke malicious beacon nodes when it determines what nodes to remove.

The Revocation Scheme

We assume each beacon node shares a unique random key with the base station. With this key, a beacon node can report its detecting results securely to the base station.

The basic idea is to evaluate the suspiciousness of each beacon node based on the alerts from detecting nodes. The beacon nodes with high degree of suspiciousness will be considered as being compromised. We measure the *suspiciousness* of a beacon node with the number of alerts against this beacon node. Since malicious beacon nodes may report many alerts against benign beacon nodes, we limit the number of alerts each beacon node can report to mitigate this effect. The detail of the algorithm is described below.

Every alert from a detecting node includes the ID of the detecting node and the ID of the target node. The base station maintains an *alert counter* and a *report counter* for each beacon node. The alert counter records the suspiciousness of this beacon node, while the report counter records the number of alerts that are reported by this node and accepted by the base station. Whenever a detecting node determines that a particular beacon node is compromised, it reports an alert to the base station. Once the base station receives the alert, it checks if the report counter of the detecting node has not exceed a threshold τ' and the target node is not revoked. If this is true, it increases both the alert counter of the target node and the report counter of the detecting node by 1; otherwise, the base station ignores this alert. The base station then checks if the alert counter of the target node exceeds another threshold τ. If yes, the target node is considered as a malicious beacon node and revoked from the network.

Note that the alert from a revoked detecting node will still be accepted by the base station if its report counter does not exceed threshold τ' and the target node is not revoked. The purpose is to prevent malicious beacon nodes from reporting a lot of alerts against benign beacon nodes and having these benign beacon nodes revoked before they can report any alert.

Analysis

For simplicity, we assume beacon nodes and non-beacon nodes in the network are randomly deployed in the field. We assume there are N sensor nodes, N_b beacon nodes, and N_a malicious beacon nodes in the network. Thus, there are $N - N_b$ non-beacon nodes and $N_b - N_a$ benign beacon nodes. We assume malicious beacon nodes do not report alerts against other malicious beacon nodes, since this will increase the probability of a malicious beacon node being detected. We also assume that every alert from beacon nodes can be successfully delivered to the base station using some standard fault tolerant techniques (e.g., retransmission) when there are message losses. When it is necessary to evaluate a certain aspect with specific numbers (e.g., in figures), we always assume 10% of sensor nodes are benign beacon nodes ($\frac{N_b - N_a}{N} = 0.1$).

Overheads: The revocation scheme requires beacon nodes to report their observations to the base station, which introduces additional communication overhead. However, a beacon node usually only needs to monitor a small number of other beacon nodes that it can communicate with. Thus, only a limited number of alerts need to be delivered to the base station. There is no additional computation overhead and storage overhead introduced by the above algorithm for the beacon nodes in the network. For the base station, it is usually not a problem to run the above algorithm, since the base station is much more resourceful than a beacon node.

Detection rate (P_d): The detection rate studied here is the probability of a malicious beacon node being revoked by the base station. Consider any requesting node u of a particular malicious beacon node v. The probability that u is a benign beacon node can be estimated by $\frac{N_b - N_a}{N}$. If node u is a benign beacon node, the probability of reporting an alert is P_r. Hence, for any requesting node, the probability of the base station having an alert reported against the malicious beacon node v can be estimated by $P_a = \frac{(N_b - N_a) \times P_r}{N}$. Suppose there are N_c requesting nodes for node v. The probability of having exactly i alerts reported can be estimated by $P(i) = \frac{N_c!}{(N_c - i)! i!} P_a^i (1 - P_a)^{N_c - i}$. Assume the threshold τ' is large enough so that an alert from a detecting node will not be ignored by the base station simply because its report counter exceeds τ'. (The method to determine τ' will be discussed later.) The probability of the number of alerts against the malicious beacon node v exceeding τ can be estimated by $P_d = 1 - \sum_{i=0}^{\tau} P(i)$, which is the expected detection rate.

Figure 5.23(a) and Figure 5.23(b) illustrate the effect of m, τ and P on the detection rate, assuming $N_c = 100$. We can see that the detection rate increases quickly when a malicious beacon node behaves maliciously more often (a larger P). In addition, the detection rate decreases with a larger threshold τ, since we need more alerts to revoke a malicious beacon node. Finally, the detection rate also increases with more detecting IDs at each

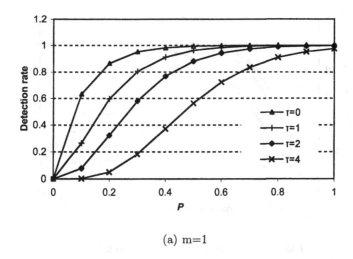

(a) m=1

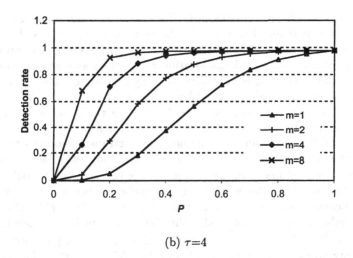

(b) τ=4

Fig. 5.23. Detection rate v.s. probability of non-beacon nodes being affected. $N_c = 100$.

beacon node, since each detecting node has more chances to detect a malicious beacon node and report an alert.

Figure 5.24 shows the effect of N_c on the detection rate, assuming $m = 8$ and $\tau = 2$. We can see that the detection rate increases when more requesting

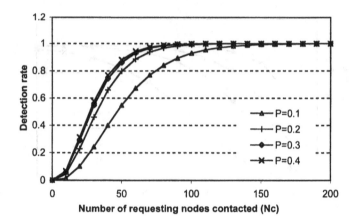

Fig. 5.24. Detection rate. $m = 8$ and $\tau = 2$.

nodes contact a malicious beacon node. This is because the more requesting nodes contact a malicious beacon node, the more alerts are reported.

Average number of affected non-beacon nodes (N'): An important target of attacks is to mislead the location estimate at as many non-beacon nodes as possible. Thus, it is necessary to study the average number of non-beacon nodes that are really affected by malicious beacon nodes. We assume that a malicious beacon signal will not be used in the location estimation if the corresponding beacon node is revoked. This can be achieved by using some standard fault tolerance techniques (e.g., retransmission) so that the revocation message from the base station can reach most of the sensor nodes.

After all detected malicious beacon nodes are revoked, the probability of a non-beacon requesting node accepting the malicious beacon signal from a malicious beacon node can be estimated by $P' = P \times (1 - P_d)$. Thus, the average number of non-beacon nodes that have been really affected can be estimated by $N' = \frac{P' \times N_c \times (N - N_b)}{N}$. Since τ and m are system parameters, the attacker may adjust P to maximize P' and thus N'. (Note that the attacker is able to control P.) Figure 5.25 shows that in practice, there are only a few non-beacon nodes accepting the malicious beacon signals. It also shows that N' as well as P' increases with a larger τ, and decreases with a larger m. This is because a malicious beacon node has a higher chance to be detected with a larger m, and a higher chance not to be revoked with a larger τ.

Figure 5.26 shows the relationship between N' and P when the attacker can always choose P to maximize N'. We can see that N' increases dramatically at the beginning. However, when N_c reaches a certain point (about 20), N' begins to drop quickly and finally remains at certain level. This is because after the number of request nodes reaches a certain point, a malicious beacon

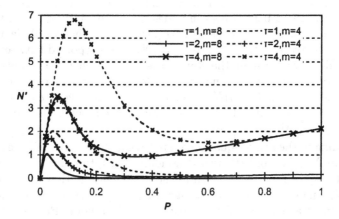

Fig. 5.25. Average number of affected non-beacon nodes after all detected malicious beacon nodes are revoked from the network. $m = 8$ and $N_c = 100$.

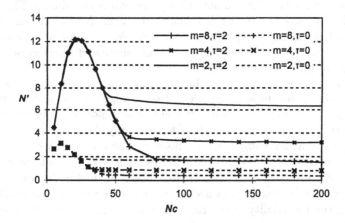

Fig. 5.26. Average number of affected non-beacon nodes when P is chosen in such a way that P' is maximized.

node has a higher chance of being revoked from the network if it is contacted by more requesting nodes. We also note that N' decreases when threshold τ decreases. This is because the probability of a malicious beacon node being revoked increases with a smaller τ.

Number of false positives (N_f): Assume there are wormhole attacks between N_w pairs of benign beacon nodes in the network. For any wormhole created between two benign beacon nodes, the probability of one of them reporting an alert against the other is $1 - p_d$, where p_d is the wormhole detection rate. Thus, on average, there are $2(1 - p_d)N_w$ alerts reported between benign beacon nodes. We consider the worst case where each beacon node reports $\tau' + 1$ alerts. Thus, the total number of alerts against benign beacon nodes can be estimated by $2(1 - p_d)N_w + N_a(\tau' + 1)$, and the average number of benign beacon nodes revoked by the base station (the number of false positives) is at most $N_f = \frac{2(1-p_d)N_w + N_a(\tau'+1)}{\tau+1}$.

According to the above equation, we note that the number of false positives depends on N_w, N_a, and the two thresholds. Thus, to reduce the number of false positives, we have to decrease τ' and/or increase τ. However, decreasing τ' implies a smaller number of alerts against a malicious node, while increasing τ implies more alerts needed to revoke a malicious node. Both of these two options will decrease the probability of malicious beacon nodes being detected. In practice, we have to make trade-offs between the number of false positives and the detection rate. The next part of the analysis will show a possible way to deal with this problem.

Thresholds τ and τ': Thresholds τ and τ' are two critical parameters. Threshold τ can be configured according to similar constraints in Figure 5.26. Intuitively, we may derive the relationship between N' and N_c as shown in Figure 5.26 given expected values of N, N_b, N_a, p_d and m. We can then choose a set of τ that makes the maximum number of affected non-beacon nodes remain under a given value.

For each of the selected thresholds τ, we configure threshold τ' in the following way so that most of the alerts from benign beacon nodes will not be ignored by the base station simply because their report counters exceed τ'.

We assume malicious nodes are also randomly deployed in the network. Consider a particular benign beacon node u. The probability of a particular malicious beacon node v being contacted by node u can be estimated by $\frac{N_c}{N}$. Since the probability of node u reporting an alert against node v is P_r and the probability of node v having not been revoked can be approximately estimated by $1 - P_d$, the probability of the report counter of node u being increased by 1 for node v can be estimated by $P_1 = \frac{P_r \times N_c \times (1 - P_d)}{N}$ if this report counter has not exceeded τ' yet. In addition, the probability of a particular wormhole being created for node u can be estimated by $\frac{2}{N_b - N_a}$, and the probability of node u reporting an alert due to this wormhole can be estimated by $1 - p_d$. Since the probability of the node at the other side of the wormhole being revoked can be approximately estimated by $\frac{N_f}{N_b - N_a}$, the probability of the report counter of node u being increased by 1 due to the wormhole attack can be estimated by $P_2 = \frac{2(1-p_d)(N_b - N_a - N_f)}{(N_b - N_a)^2}$ if this report counter has not exceeded τ' yet. Hence, the probability that the report counter of node u is i $(i \leq \tau')$ can be estimated by

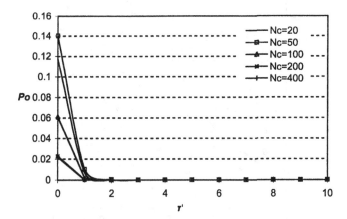

Fig. 5.27. Probability of the report counter of a benign beacon node exceeding τ'. Assume $N = 10,000$, $N_b = 1100$, $N_a = 100$, $N_w = 100$, $p_d = 0.9$, $\tau = 2$, $m = 8$, and $P = 0.1$.

$$P'(i) = \sum_{j+k=i} \frac{N_a! N_w! P_1^j (1-P_1)^{N_a-j} P_2^k (1-P_2)^{N_w-k}}{(N_a-j)! j! (N_w-k)! k!}.$$

Therefore, the probability of the report counter of a benign node exceeding τ' can be estimated by $P_o = 1 - \sum_{i=0}^{\tau'} P'(i)$. Figure 5.27 plots this probability when $\tau = 2$, assuming $N = 10,000$, $N_b = 1100$, $N_a = 100$, $N_w = 100$, $p_d = 0.9$, $m = 8$, and $P = 0.1$. We can see that the probability of the report counter of a benign beacon node exceeding 2 is close to zero. Thus, we can chose $\tau' = 2$ and have a pair of candidate thresholds ($\tau = 2, \tau' = 2$). We also note that malicious beacon nodes cannot increase this probability by simply having more requesting nodes contact it, since this will increase the chance of being revoked.

After the above analysis, we can find a proper threshold τ' for each selected τ. We then choose a pair of thresholds τ and τ' that satisfy the constraints on the number of false positives N_f or simply choose a pair of thresholds that lead to the minimum N_f given certain p_d, N_w and N_a.

5.4.4 Simulation Evaluation

We have implemented the proposed techniques on TinyOS [26], an operation system for networked sensors. In this section, we present the results obtained through the TinyOS simulator Nido, with a focus on the detection rate and the false positive rate (i.e., $\frac{\#incorrect\ revoked\ beacons}{\#total\ benign\ beacons}$) of the proposed schemes.

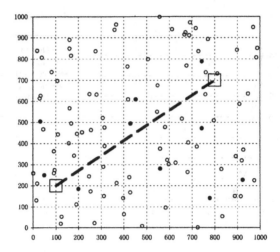

Fig. 5.28. Deployment of beacon nodes in a sensing field.

We assume 1,000 sensor nodes ($N = 1000$) randomly deployed in a sensing field of 1000×1000 square feet. Among these sensor nodes, there are 100 beacon nodes ($N_b = 100$) with 10 compromised beacon nodes ($N_a = 10$). Figure 5.28 shows the randomly generated deployment used in our simulation, where each blank circle (○) represents a benign beacon node and each solid circle (●) represents a malicious beacon node. We assume the maximum communication range of a beacon or non-beacon node is 150 feet and a malicious beacon node only contacts the nodes within its communication range.

In the simulation there is a wormhole between location A (100,200) and location B (800,700), which forwards every message received at one side immediately to the other side. We assume malicious beacon nodes collude together to report alerts against benign beacon nodes. Thus, they can always make the base station revoke about $\frac{N_a \times (\tau' + 1)}{\tau + 1}$ benign beacon nodes by simply reporting alerts. We always assume $m = 8$ and $p_d = 0.9$. We also assume that there is a technique (e.g., RSSI) used to estimate the distance to the beacon node that has the maximum error of 10 feet.

Figure 5.29 shows the detection rate when $\tau' = 2$ and $\tau = 2$. The result conforms to the theoretical analysis. We can clearly see the increase in the detection rate when a malicious beacon node tries to increase P to affect more non-beacon nodes. Figure 5.30 shows the average number N' of requesting non-beacon nodes accepting the malicious beacon signals from a malicious beacon node. We note that the simulation result has observable but small difference from the theoretical analysis. The simulation result and the theoretical result are in general close to each other.

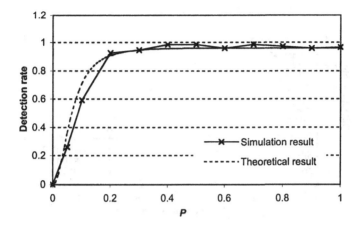

Fig. 5.29. Detection rate v.s. P. Assume $\tau' = 2$ and $\tau = 2$.

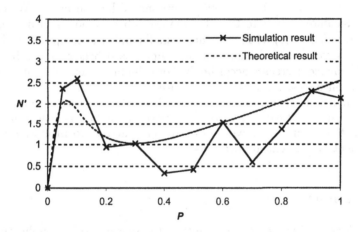

Fig. 5.30. Average number of requesting non-beacon nodes accepting the malicious beacon signals from a malicious beacon node. Assume $\tau' = 2$ and $\tau = 2$.

Based on our earlier analysis, both the detection rate and the false positive rate are affected by τ and τ' given certain p_d, N_w and N_a. Figure 5.31 shows the ROC (Receiver Operating Characteristic) curves for the proposed techniques under different choice of τ and τ', assuming P is configured in such a way that N' is maximized. (The various points in the figure are obtained by using different values of τ.) It includes the performance when there are

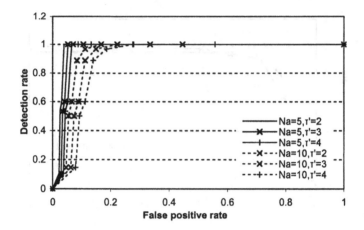

Fig. 5.31. ROC curves. Assume P is chosen to maximize N'.

either 5 ($N_a = 5$) or 10 ($N_a = 10$) compromised beacon nodes. We can see that our technique can detect most of malicious beacon nodes with small false positive rate (e.g., 5%) when there are a small number of compromised beacon nodes. However, when the number of compromised beacon nodes increases, the performance decreases accordingly. For example, when there are 10 malicious beacon nodes, the false positive rate will reach 20% in order to detect most of malicious beacon nodes. Nevertheless, the figure still shows that our techniques are practical and effective in detecting malicious beacon nodes. In addition, this figure also gives a way to set τ and τ' to meet the security requirement of different applications.

5.5 Summary

Sensors' locations are critical to many wireless sensor network applications. In this chapter, we identified the vulnerabilities of current localization schemes for sensor networks in hostile environments where there are malicious attacks. To deal with such attacks, we proposed an attack-resistant MMSE-based location estimation and a voting-based location estimation technique. When beacon packets are authenticated and beacon nodes are uniquely identified, these techniques substantially improve the security and the robustness of location estimation in wireless sensor networks. We also developed a number of techniques to detect malicious beacon nodes that supply incorrect information for location discovery. Our experiences indicate that the proposed techniques are practical solutions for securing location discovery in wireless sensor networks.

Our future research is two-fold. First, we will study alternative and potentially more efficient and effective ways for secure localization in wireless sensor networks. Second, we will study how to combine the proposed techniques with other protection mechanisms such as wormhole detection.

6

Summary and Future Work

Wireless sensor networks have received a lot of attention recently due to the wide range of potential applications in civilian and military operations. Security becomes one of the main concerns in hostile environments. However, some unique features of these networks, such as node compromises and resource constraints, make traditional security mechanisms not as effective as in wired networks. This book focuses on security techniques for critical components in wireless sensor networks.

6.1 Summary

In this book, we first present two efficient techniques for broadcast authentication in sensor networks. The multi-level μTESLA extends the capability of the original μTESLA by removing the dependency on unicast-based distribution for the initial commitments. The tree-based μTESLA further improves the original μTESLA in the sense that (1) it allows broadcast authentication for a large number of senders and (2) it is not subject to the DoS attacks. These two methods extended the application of μTESLA in various situations.

Pairwise key establishment enables sensor nodes to communicate securely with each other using cryptographic techniques. We developed a number of key pre-distribution techniques to deal with this problem [42, 48]. We first presented a general framework for pairwise key establishment based on the polynomial-based key pre-distribution [6] and the probabilistic key distribution [20]. By instantiating the components in this framework, we further developed two novel pairwise key pre-distribution schemes: a random subset assignment scheme and a hypercube-based scheme. Both of them can achieve better performance than the previous methods. In addition, we also studied how to take advantage of the prior deployment knowledge, post deployment knowledge and group-based deployment knowledge of sensor nodes to improve the performance of existing key pre-distribution techniques [43, 44].

Sensors' locations are of particular importance in many sensor network applications. Protecting localization is particularly challenging in the presence of malicious attacks. We first developed two methods to survive malicious attacks against the location discovery in sensor networks. The first method filters out malicious beacon signals on the basis of the "consistency" among multiple beacon signals, while the second method tolerates malicious beacon signals by adopting an iteratively refined voting scheme. Both methods can survive malicious attacks even if the attackers bypass traditional cryptographic protections such as authentication, as long as the benign beacon signals constitute the majority of the "consistent" beacon signals. In addition, we also proposed a suite of techniques to detect and remove the compromised beacon nodes that supply misleading location information to regular sensor nodes. These techniques further improve the security of location discovery in sensor networks.

6.2 Future Work

Despite the substantial advances in techniques for securing wireless sensor networks, many security problems still haven't been fully addressed and still need further investigation.

1. *Fundamental cryptographic mechanisms*: It is worth further studying those fundamental cryptographic mechanisms in wireless sensor networks. One example could be broadcast authentication. Though the μTESLA protocol removes the dependency on public key cryptography for broadcast authentication in sensor networks, it requires loose time synchronization between a sender and multiple receivers. Hence, providing a practical broadcast authentication protocol without depending on time synchronization is of particular interest for sensor networks. In addition, some other problems such as group key establishment, key update and key revocation also need further investigation.

   On the other hand, several recent experiments show that sensor nodes are able to compute a few optimized public key operations [3]. This leads to an interesting research direction: how to provide efficient broadcast authentication protocols using optimized public key cryptography. In particular, we need to further reduce the storage and communication overheads and mitigate the DOS attacks introduced by the public key operations.

2. *Security of fundamental services*: In our earlier studies, we only looked at the problem of secure location discovery. However, there are many other fundamental services that need protection, for example, data management, routing, and secure time synchronization. Moreover, secure localization is still a challenging problem in wireless sensor networks. We believe that there is still room for improving the resilience of current techniques. For example, we may use additional knowledge such as topology

information to further improve the detection and revocation of compromised beacon nodes.

3. *Detection of attacks*: Many attacks have been identified in sensor networks; for example, node capture attacks, sybil attacks, and wormhole attacks. Due to the resource constraints on sensor nodes and node capture attacks, the detection of attacks in sensor networks is different and potentially more difficult than the intrusion detection in traditional networks. An important question that we need to address is how to distinguish malicious behavior from normal behavior. One possible approach is to use the application semantics. In addition, we also need to study how to conduct such detection in a distributed and localized fashion and how to cooperate with each other to improve the detection result.

References

1. I.F. Akyildiz, W. Su, Y. Sankarasubramaniam, and E. Cayirci. Wireless sensor networks: A survey. *Computer Networks*, 38(4):393–422, 2002.
2. R. Anderson, F. Bergadano, B. Crispo, J. Lee, C. Manifavas, and R. Needham. A new family of authentication protocols. In *Operating Systems Review*, October 1998.
3. N. Gura, A. Patel, and A. Wander. Comparing Elliptic Curve Cryptography and RSA on 8-bit CPUs. in *Proceedings of the 2004 Workshop on Cryptographic Hardware and Embedded Systems (CHES 2004)*, August 2004.
4. S. Basagni, K. Herrin, D. Bruschi, and E. Rosti. Secure pebblenets. In *Proceedings of ACM International Symposium on Mobile ad hoc networking and computing*, pages 156–163, 2001.
5. F. Bergadano, D. Cavagnino, and B. Crispo. Individual single source authentication on the mbone. In *IEEE International Conference on Multimedia & Expo (ICME)*, August 2000.
6. C. Blundo, A. De Santis, Amir Herzberg, S. Kutten, U. Vaccaro, and M. Yung. Perfectly-secure key distribution for dynamic conferences. In *Advances in Cryptology – CRYPTO '92, LNCS 740*, pages 471–486, 1993.
7. B. Briscoe. FLAMeS: Fast, loss-tolerant authentication of multicast stream. Technical report, BT Research, 2000.
8. S. Buchegger and J. L. Boudec. Performance analysis of the CONFIDANT protocol (cooperation of nodes: Fairness in dynamic ad-hoc networks). In *Proceedings of The Third ACM International Symposium on Mobile Ad Hoc Networking and Computing*, pages 226–236, June 2002.
9. N. Bulusu, J. Heidemann, and D. Estrin. GPS-less low cost outdoor localization for very small devices. In *IEEE Personal Communications Magazine*, pages 28–34, October 2000.
10. R. Canetti, J. Garay, G. Itkis, D. Micciancio, M. Naor, and B. Pinkas. Multicast security: A taxonomy and some efficient constructions. In *Proceedings of IEEE INFOCOM '99*, pages 708–716, 1999.
11. D.W. Carman, P.S. Kruus, and B.J.Matt. Constrains and approaches for distributed sensor network security. Technical report, NAI Labs, 2000.
12. H. Chan, A. Perrig, and D. Song. Random key predistribution schemes for sensor networks. In *IEEE Symposium on Research in Security and Privacy*, pages 197–213, 2003.

13. S. Cheung. An efficient message authentication scheme for link state routing. In *13th Annual Computer Security Applications conference*, San Diego, California, December 1997.

14. Crossbow Technology Inc. Wireless sensor networks. `http://www.xbow.com/ Products/Wireless_Sensor_Networks.htm`. Accessed in May 2005.

15. J. Deng, R. Han, and S. Mishra. Security support for in-network processing in wireless sensor networks. In *2003 ACM Workshop on Security in Ad Hoc and Sensor Networks (SASN '03)*, October 2003.

16. L. Doherty, K. S. Pister, and L. E. Ghaoui. Convex optimization methods for sensor node position estimation. In *Proceedings of INFOCOM'01*, 2001.

17. W. Du, J. Deng, Y. S. Han, S. Chen, and P. Varshney. A key management scheme for wireless sensor networks using deployment knowledge. In *Proceedings of IEEE INFOCOM'04*, March 2004.

18. W. Du, J. Deng, Y. S. Han, and P. Varshney. A pairwise key pre-distribution scheme for wireless sensor networks. In *Proceedings of 10th ACM Conference on Computer and Communications Security (CCS'03)*, pages 42–51, October 2003.

19. W. Du, L. Fang, and P. Ning. Lad: Localization anomaly detection for wireless sensor networks. In *Proceedings of the 19th IEEE International Parallel & Distributed Processing Symposium (IPDPS '05)*, April 2005.

20. L. Eschenauer and V. D. Gligor. A key-management scheme for distributed sensor networks. In *Proceedings of the 9th ACM Conference on Computer and Communications Security*, pages 41–47, November 2002.

21. D. Gay, P. Levis, R. von Behren, M. Welsh, E. Brewer, and D. Culler. The nesC language: A holistic approach to networked embedded systems. In *Proceedings of Programming Language Design and Implementation (PLDI 2003)*, June 2003.

22. R. Gennaro and P. Rohatgi. How to sign digital streams. Technical report, IBM T.J.Watson Research Center, 1997.

23. O. Goldreich, S. Goldwasser, and S. Micali. How to construct random functions. *Journal of the ACM*, 33(4):792–807, October 1986.

24. N. M. Haller. The S/KEY one-time password system. In *Proceedings of the ISOC Symposium on Network and Distributed System Security*, pages 151–157, 1994.

25. T. He, C. Huang, B. M. Blum, J. A. Stankovic, and T. F. Abdelzaher. Range-free localization schemes in large scale sensor networks. In *Proceedings of ACM MobiCom 2003*, 2003.

26. J. Hill, R. Szewczyk, A. Woo, S. Hollar, D.E. Culler, and K. S. J. Pister. System architecture directions for networked sensors. In *Architectural Support for Programming Languages and Operating Systems*, pages 93–104, 2000.

27. L. Hu and D. Evans. Secure aggregation for wireless networks. In *Workshop on Security and Assurance in Ad Hoc Networks*, January 2003.

28. L. Hu and D. Evans. Using directional antennas to prevent wormhole attacks. In *Proceedings of the 11th Network and Distributed System Security Symposium*, pages 131–141, February 2003.

29. Y.C. Hu, A. Perrig, and D.B. Johnson. Packet leashes: A defense against wormhole attacks in wireless ad hoc networks. In *Proceedings of INFOCOM 2003*, April 2003.

30. D. Huang, M. Mehta, D. Medhi, and L. Harn. Location-aware key management scheme for wireless sensor networks. In *Proceedings of the 2nd ACM workshop*

on Security of ad hoc and sensor networks (SASN '04), pages 29 – 42, October 2004.

31. C. Intanagonwiwat, R. Govindan, and D. Estrin. Directed diffusion: A scalable and robust communication paradigm for sensor networks. In *Proceedings of the sixth annual international conference on Mobile computing and networking (Mobicom '00)*, pages 56–67, Nov 2003.

32. C. Karlof, N. Sastry, Y. Li, A. Perrig, and J. Tygar. Distillation codes and applications to dos resistant multicast authentication. In *Proceedings of the 11th Network and Distributed Systems Security Symposium (NDSS)*, 2004.

33. C. Karlof, N. Sastry, and D. Wagner. TinySec: Link layer encryption for tiny devices. http://www.cs.berkeley.edu/~nks/tinysec/.

34. C. Karlof and D. Wagner. Secure routing in wireless sensor networks: Attacks and countermeasures. In *Proceedings of 1st IEEE International Workshop on Sensor Network Protocols and Applications*, May 2003.

35. B. Karp and H. T. Kung. GPSR: Greedy perimeter stateless routing for wireless networks. In *Proceedings of ACM MobiCom 2000*, 2000.

36. D.E. Knuth. *The Art of Computer Programming*, volume 2: Seminumerical Algorithms. Addison-Wesley, third edition, 1997. ISBN: 0-201-89684-2.

37. H. Krawczyk, M. Bellare, and R. Canetti. HMAC: Keyed-hashing for message authentication. IETF RFC 2104, February 1997.

38. L. Lamport. Password authentication with insecure communication. *Communications of the ACM*, 24(11):770–772, 1981.

39. L. Lazos and R. Poovendran. Serloc: Secure range-independent localization for wireless sensor networks. In *ACM workshop on Wireless security (ACM WiSe 2004)*, Philadelphia, PA, October 1 2004.

40. L. Li and J.Y. Halpern. Minimum-energy mobile wireless networks revisited. In *Proceedings of IEEE International Conference on Communications (ICC '01)*, June 2001.

41. D. Liu and P. Ning. Efficient distribution of key chain commitments for broadcast authentication in distributed sensor networks. In *Proceedings of the 10th Annual Network and Distributed System Security Symposium (NDSS'03)*, pages 263–276, February 2003.

42. D. Liu and P. Ning. Establishing pairwise keys in distributed sensor networks. In *Proceedings of 10th ACM Conference on Computer and Communications Security (CCS'03)*, pages 52–61, October 2003.

43. D. Liu and P. Ning. Location-based pairwise key establishments for static sensor networks. In *2003 ACM Workshop on Security in Ad Hoc and Sensor Networks (SASN '03)*, pages 72–82, October 2003.

44. D. Liu and P. Ning. Improving key pre-distribution with deployment knowledge in static sensor networks. Submitted for publication, 2004.

45. D. Liu and P. Ning. Multi-level μTESLA: Broadcast authentication for distributed sensor networks. *ACM Transactions in Embedded Computing Systems (TECS)*, 3(4), 2004.

46. D. Liu, P. Ning, and W.K. Du. Attack-resistant location estimation in wireless sensor networks. In *Proceedings of the Fourth International Conference on Information Processing in Sensor Networks (IPSN '05)*, April 2005.

47. D. Liu, P. Ning, and W.K. Du. Detecting malicious beacon nodes for secure location discovery in wireless sensor networks. In *Proceedings of the 25th International Conference on Distributed Computing Systems (ICDCS '05)*, June 2005.

48. D. Liu, P. Ning, and R. Li. Establishing pairwise keys in distributed sensor networks. *ACM Transactions on Information and System Security (TISSEC)*, 2004. To appear.

49. S. Marti, T. J. Giuli, K. Lai, and M. Baker. Mitigating routing misbehavior in mobile ad hoc networks. In *Proceedings of the Sixth annual ACM/IEEE International Conference on Mobile Computing and Networking*, pages 255–265, 2000.

50. R. Merkle. Protocols for public key cryptosystems. In *Proceedings of the IEEE Symposium on Research in Security and Privacy*, Apr 1980.

51. R. Nagpal, H. Shrobe, and J. Bachrach. Organizing a global coordinate system from local information on an ad hoc sensor network. In *IPSN'03*, 2003.

52. A. Nasipuri and K. Li. A directionality based location discovery scheme for wireless sensor networks. In *Proceedings of ACM WSNA'02*, September 2002.

53. J. Newsome, R. Shi, D. Song, and A. Perrig. The sybil attack in sensor networks: Analysis and defenses. In *Proceedings of IEEE International Conference on Information Processing in Sensor Networks (IPSN 2004)*, Apr 2004.

54. J. Newsome and D. Song. GEM: graph embedding for routing and data-centric storage in sensor networks without geographic information. In *Proceedings of the First ACM Conference on Embedded Networked Sensor Systems (SenSys '03)*, pages 76–88, Nov 2003.

55. D. Niculescu and B. Nath. Ad hoc positioning system (APS). In *Proceedings of IEEE GLOBECOM '01*, 2001.

56. D. Niculescu and B. Nath. Ad hoc positioning system (APS) using AoA. In *Proceedings of IEEE INFOCOM 2003*, pages 1734–1743, April 2003.

57. D. Niculescu and B. Nath. DV based positioning in ad hoc networks. In *Journal of Telecommunication Systems*, 2003.

58. NIST. Skipjack and KEA algorithm specifications. `http://csrc.nist.gov/encryption/skipjack/skipjack.pdf`, May 1998.

59. A. Perrig. The BiBa one-time signature and broadcast authentication protocol. In *Proceedings of the ACM Conference on Computer and Communications Security*, pages 28–37, November 2001.

60. A. Perrig, R. Canetti, Briscoe, J. Tygar, and D. Song. TESLA: Multicast source authentication transform. IRTF draft, draft-irtf-smug-tesla-00.txt, November 2000.

61. A. Perrig, R. Canetti, D. Song, and D. Tygar. Efficient authentication and signing of multicast streams over lossy channels. In *Proceedings of the 2000 IEEE Symposium on Security and Privacy*, May 2000.

62. A. Perrig, R. Canetti, D. Song, and D. Tygar. Efficient and secure source authentication for multicast. In *Proceedings of Network and Distributed System Security Symposium*, February 2001.

63. A. Perrig, R. Szewczyk, V. Wen, D. Culler, and D. Tygar. SPINS: Security protocols for sensor networks. In *Proceedings of Seventh Annual International Conference on Mobile Computing and Networks*, July 2001.

64. R. D. Pietro, L. V. Mancini, and A. Mei. Random key assignment for secure wireless sensor networks. In *2003 ACM Workshop on Security in Ad Hoc and Sensor Networks (SASN '03)*, October 2003.

65. B. Przydatek, D. Song, and A. Perrig. SIA: Secure information aggregation in sensor networks. In *Proceedings of the First ACM Conference on Embedded Networked Sensor Systems (SenSys '03)*, Nov 2003.

66. S. Ratnasamy, B. Karp, L. Yin, F. Yu, D. Estrin, R. Govindan, and S. Shenker. GHT: A geographic hash table for data-centric storage. In *Proceedings of 1st ACM International Workshop on Wireless Sensor Networks and Applications*, Sep 2002.

67. R. Rivest. The RC5 encryption algorithm. In *Proceedings of the 1st International Workshop on Fast Software Encryption*, volume 809, pages 86–96, 1994.

68. R.L. Rivest, A. Shamir, and L.A. Adleman. A method for obtaining digital signatures and public-key cryptosystems. *Communications of the ACM*, 21(2):120–126, 1978.

69. P. Rohatgi. A compact and fast hybrid signature scheme for multicast packet authentication. In *6th ACM Conference on Computer and Communications Security*, November 1999.

70. N. Sastry, U. Shankar, and D. Wagner. Secure verification of location claims. In *ACM Workshop on Wireless Security*, 2003.

71. A. Savvides, C. Han, and M. Srivastava. Dynamic fine-grained localization in ad-hoc networks of sensors. In *Proceedings of ACM MobiCom '01*, pages 166–179, July 2001.

72. A. Savvides, H. Park, and M. Srivastava. The bits and flops of the n-hop multilateration primitive for node localization problems. In *Proceedings of ACM WSNA '02*, September 2002.

73. S. Shenker, S. Ratnasamy, B. Karp, R. Govindan, and D. Estrin. Data-centric storage in sensornets. In *Proceedings of the First ACM Workshop on Hot Topics in Networks*, October 2002.

74. V. Shnayder, M. Hempstead, B. Chen, G. Werner-Allen, and M. Welsh. Simulating the power consumption of large-scale sensor network applications. In *Proceedings of the Second ACM Conference on Embedded Networked Sensor Systems (SenSys'04)*, Nov 2004.

75. F. Stajano and R. Anderson. The resurrecting duckling: security issues for ad hoc networks. In *Proceedings of the 7th International Workshop on Security Protocols*, pages 172–194, 1999.

76. W. Stallings. *Cryptography and Network Security: Principles and Practice*. Prentice Hall, 2nd edition, 1999.

77. C.K. Wong and S. S. Lam. Digital signatures for flows and multicasts. In *Proc. IEEE ICNP'98*, 1998.

78. D. Wong and A. Chan. Efficient and mutually authenticated key exchange for low power computing devices. In *Proceedings of ASIA CRYPT*, Dec 2001.

79. A. D. Wood and J. A. Stankovic. Denial of service in sensor networks. *IEEE Computer*, 35(10):54–62, 2002.

80. Y. Yu, R. Govindan, and D. Estrin. Geographical and energy aware routing: A recursive data dissemination protocol for wireless sensor networks. Technical Report UCLA/CSD-TR-01-0023, UCLA, Department of Computer Science, May 2001.

81. S. Zhu, S. Setia, and S. Jajodia. LEAP: Efficient security mechanisms for large-scale distributed sensor networks. In *Proceedings of 10th ACM Conference on Computer and Communications Security (CCS'03)*, pages 62–72, October 2003.

82. S. Zhu, S. Setia, S. Jajodia, and P. Ning. An interleaved hop-by-hop authentication scheme for filtering false data in sensor networks. In *Proceedings of 2004 IEEE Symposium on Security and Privacy*, May 2004.

Index